Animal Assisted Therapy
USE APPLICATION BY CONDITION

Animal Assisted Therapy USE APPLICATION BY CONDITION

Edited by

ERIC ALTSCHULER

Department of Physical Medicine and Rehabilitation, Metropolitan Hospital, New York, NY, United States;

Department of Rehabilitation Medicine, New York Medical College, Valhalla, NY, United States

ELSEVIER

ACADEMIC PRESS

An imprint of Elsevier

Academic Press is an imprint of Elsevier
125 London Wall, London EC2Y 5AS, United Kingdom
525 B Street, Suite 1650, San Diego, CA 92101, United States
50 Hampshire Street, 5th Floor, Cambridge, MA 02139, United States
The Boulevard, Langford Lane, Kidlington, Oxford OX5 1GB, United Kingdom

ISBN: 978-0-323-98815-5

For information on all Academic Press publications visit our website at https://www.elsevier.com/books-and-journals

Publisher: Stacy Masucci
Acquisitions Editor: Nikki P. Levy
Editorial Project Manager: Anna Valutkevich
Production Project Manager: Niranjan Bhaskaran
Cover Designer: Mark Rogers

Typeset by TNQ Technologies

Dedication

We dedicate the book to our heroes and heroines, the unseen heroes and heroines who fight PTSD on a daily basis and to the extraordinary service animals that help these heroes and heroines stay healthy.

Romeo and Juliet (1597 Quarto), Act I, Scene 4

And then dreames he of cutting forraine throats,
Of breaches ambuscados, countermines,
Of healthes fiue fadome deepe, and then anon
Drums in his eare: at which he startes and wakes,
And sweares a Praier or two and sleepes againe.

Contents

List of contributors *xv*
Acknowledgments *xvii*

SECTION 1 Introduction

1. Introduction and synopsis of chapters **3**
Eric Altschuler

References 7

2. History of animal-assisted therapy (AAT) and theories of mechanism of action of AAT for posttraumatic stress disorder (PTSD) **9**
Eric Altschuler

References 14

3. Incorporating animal-assisted therapy into treatment for posttraumatic stress disorder **17**
Carin M. Lefkowitz

Introduction 17
The benefits of animal-assisted therapy for survivors of trauma 18
The state of the research 20
A brief summary of evidence-based psychotherapies for PTSD 22
Models for incorporating AAT into evidence-based psychotherapies for PTSD 23
Case examples 27
Phil and Fonzie: the dog that aided exposure therapy 27
Conclusions 30
References 30

4. Posttraumatic stress disorder in history, literature, and art **33**
Eric Altschuler

References 40

SECTION 2 Applications of AAT

5. The use of service dogs for military veterans with posttraumatic stress disorder **45**
James Whitworth

Introduction 45
Description of service dogs for veterans with PTSD 46
Service dog training programs for veterans with PTSD 47
Current guidelines for selecting and pairing service dogs for veterans 49
Reported benefits of service dogs for veterans with PTSD 50
Mental health benefits 51
Improvements in relational/social/occupational functioning 54
Physical/medical impacts 54
Impacts of service dogs on family members of veterans with PTSD 55
US Department of Veterans Affairs study of service dogs for veterans with PTSD 56
Challenges in studying the use of service dogs for veterans with PTSD 56
Theoretical underpinnings of service dog programs 58
Conclusion 59
References 60

6. Equine-assisted therapy for veterans with PTSD **65**
Diane Scotland-Coogan

Introduction 65
Military culture and treatment 66
Posttraumatic stress disorder 66
Experiential learning 72
Ethical practices 73
Equine therapy with veterans 73
Family considerations 74
Substance abuse disorder and equine therapy 76
What does this mean for veterans? 76
Independence and self-mastery 77
Focus, concentration, and memory 77
Relationships 78
Anxiety, depression, and anger 78
References 78

7. Avian-assisted therapy for posttraumatic stress disorder **85**
Greg Para

Why parrots? 86
My story 88
How new is avian therapy? 91
Establishing an avian therapy program 92
Creating an avian environment 93
References 95

8. Animal-assisted therapy for posttraumatic stress disorder in sexual trauma survivors **97**
Simone Swartzentuber Emmons

Personal story 97
Benefits of service dogs for survivors of sexual assault diagnosed with PTSD 102
How to approach a survivor of sexual assault and service dog team in a medical setting? 107
Service Dog Strong organization 111
Potential drawbacks to having a service dog 115
Conclusion 116
References 117

9. Animal-assisted therapy for pediatric patients **119**
Beth L. Macauley

Dogs 120
Horses 129
Other animals 135
Conclusion 137
References 137

10. Animal-assisted therapy for cardiac conditions **147**
Sami V. Abate

Prevalence of heart disease and heart failure 147
Heart failure pathophysiology and management 147
Animal-assisted interventions and cardiac outcomes 148
Animal-assisted interventions and heart rate modulation 149
Animal-assisted interventions and stress mitigation 151
Animal-assisted interventions and physical activity 153

Exemplar: the application of animal-assisted interventions in heart failure 155
Animal-assisted interventions and cardiac outcomes in unique populations 158
Looking back and forward: the heart of animal-assisted interventions 159
References 161

11. Animal-assisted therapy in geriatric patients **165**
Dominic M. DeMarco, Kala Chinnaswamy and George T. Grossberg

References 168

SECTION 3 Animal trainers' and patient perspectives

12. Canine-assisted therapy for posttraumatic stress disorder in war veterans: perspective of an animal training organization **173**
Jessica Nelson

Mission, structure, and overview of Paws for Heroes 173
Paws for Heroes veteran clients 174
Application process 176
Ineligible veterans 177
In-person interview meeting 177
Service animals versus companion animals 178
The legal difference between a service animal and a companion animal 179
In-home evaluation 181
Paws for Heroes' dog evaluation and placement process 181
Assessment process 181
Veterinarian wellness exam 187
Foster process 188
Providing professional dog training 188
Presenting the dog to the veteran 188
Providing additional training to the veteran and their emotional support dog 189
"After-care" financial support 189
References 190

13. Animal welfare **191**
Chalotte Glintborg

Introduction 191
Mental illness and animal-assisted interventions 192
Training methods and link between animal abuse and violence toward humans 195

Canine body language 196
Suggested guidelines 206
Conclusions 210
References 210

14. Canine-assisted therapy for posttraumatic stress disorder: a war veteran's perspective 215
Paws for Heroes Client - Army Veteran

SECTION 4 Conclusion

15. Conclusion and future studies and applications of animal-assisted therapy 221
Eric Altschuler

References 223

Index 225

List of contributors

Sami V. Abate
Inspira Health, AVP Clinical Research and Nursing Excellence, Vineland, NJ, United States

Eric Altschuler
Department of Physical Medicine and Rehabilitation, Metropolitan Hospital, New York, NY, United States; Department of Rehabilitation Medicine, New York Medical College, Valhalla, NY, United States

Kala Chinnaswamy
Saint Louis University School of Medicine, Department of Psychiatry and Behavioral Neuroscience, Saint Louis, MO, United States

Dominic M. DeMarco
Saint Louis University School of Medicine, Department of Psychiatry and Behavioral Neuroscience, Saint Louis, MO, United States

Simone Swartzentuber Emmons
Service Dog Strong, Portland, ME, United States

Chalotte Glintborg
Aalborg University, Department of Communication and Psychology, Aalborg, Denmark

George T. Grossberg
Saint Louis University School of Medicine, Division of Geriatric Psychiatry, Department of Psychiatry and Behavioral Neuroscience, Saint Louis, MO, United States

Carin M. Lefkowitz
Private Practice, Philadelphia, PA, United States

Beth L. Macauley
Department of Communication Sciences and Disorders, Grand Valley State University, Grand Rapids, MI, United States

Jessica Nelson
Paws for Heroes, Houston, TX, United States

Greg Para
Sarasota Parrot Conservatory, Myakka, FL, United States

Paws for Heroes Client - Army Veteran
Houston, TX, United States

Diane Scotland-Coogan
School of Education and Social Services, Saint Leo University, Saint Leo, FL, United States

James Whitworth
University of Central Florida, Orlando, FL, United States

Acknowledgments

Thanks to Anna Valutkevich from Elsevier for suggesting writing a comprehensive volume on animal-assisted therapy and to Kyle Gravel and the rest of the editorial and production teams from Elsevier for helping bring Anna's idea to fruition. The old saying is that the secret to being seen as a good book editor is to have good chapter authors. I am fortunate to have chapter authors who make me look like a great editor. I especially appreciate that all of the chapter authors are not only scholars in their respective fields and disciplines, but also front-line practitioners of the same. I thank Sheila Brody and Dr. Hillel Swiller for helpful discussions.

Eric L Altschuler, MD, PhD

SECTION 1

Introduction

CHAPTER 1

Introduction and synopsis of chapters

Eric Altschuler[1,2]
[1]Department of Physical Medicine and Rehabilitation, Metropolitan Hospital, New York, NY, United States; [2]Department of Rehabilitation Medicine, New York Medical College, Valhalla, NY, United States

Posttraumatic stress disorder (PTSD, see Table 1.1 for diagnostic criteria for PTSD) can follow war trauma, sexual abuse, or other traumas (Bisson et al., 2015; Shalev et al., 2017). PTSD symptoms can also be experienced by caregivers of individuals with PTSD (Clawson et al., 2013), and even military commanders can experience with PTSD responses regarding trauma experienced by their subordinates (Altschuler, 2016). Medications and counseling are often not effective in treating PTSD (Shalev et al., 2017; Raskind et al., 2018), so new treatments are needed. Some years ago, I suggested (Altschuler, 1999) that animal-assisted therapy (AAT) might be beneficial for PTSD. Since then, use of AAT for PTSD has grown dramatically and is now used worldwide.

In this book, experts in their respective fields, all of whom are also practitioners, first discuss and then review the theoretical basis and incorporation of AAT in the treatment of PTSD. The book includes a chapter on PTSD in history, literature, and art, which is followed by chapters outlining the evidence for the use of canine and equine-assisted therapy for PTSD and a chapter on the use of avian-assisted therapy for individuals with PTSD. Subsequent chapters address the use and implementation of AAT for PTSD in sexual trauma survivors, and use of AAT in pediatric, geriatric, and heart failure patients. Finally, the process of setting up an organization to provide emotional support canines is covered along with a review of the important issue of animal welfare. The book concludes with the powerful narrative of a war veteran's benefit from AAT and a return in light of this to a soliloquy from Shakespeare's Henry IV, Part 1.

Following this introduction and synopsis, in Chapter 2, I review the history of AAT and its use to treat PTSD and discuss some theories on the mechanism of AAT in PTSD.

Animal Assisted Therapy Use Application by Condition
ISBN 978-0-323-98815-5
https://doi.org/10.1016/B978-0-323-98815-5.00004-5

Table 1.1 Diagnostic criteria for posttraumatic stress disorder.

Note: The following criteria apply to adults, adolescents, and children older than six years. For children six years and younger, see corresponding criteria below.

A. Exposure to actual or threatened death, serious injury, or sexual violence in one (or more) of the following ways:
   1. Directly experiencing the traumatic event(s).
   2. Witnessing, in person, the event(s) as it occurred to others.
   3. Learning that the traumatic event(s) occurred to a close family member or close friend. In cases of actual or threatened death of a family member or friend, the event(s) must have been violent or accidental.
   4. Experiencing repeated or extreme exposure to aversive details of the traumatic event(s) (e.g., first responders collecting human remains; police officers repeatedly exposed to details of child abuse).

   Note: Criterion A4 does not apply to exposure through electronic media, television, movies, or pictures, unless this exposure is work related.

B. Presence of one (or more) of the following intrusion symptoms associated with the traumatic event(s), beginning after the traumatic event(s) occurred:
   1. Recurrent, involuntary, and intrusive distressing memories of the traumatic event(s).

   Note: In children older than six years, repetitive play may occur in which themes or aspects of the traumatic event(s) are expressed.

   2. Recurrent distressing dreams in which the content and/or affect of the dream are related to the traumatic event(s).

   Note: In children, there may be frightening dreams without recognizable content.

   3. Dissociative reactions (e.g., flashbacks) in which the individual feels or acts as if the traumatic event(s) were recurring. (Such reactions may occur on a continuum, with the most extreme expression being a complete loss of awareness of present surroundings).

   Note: In children, trauma-specific reenactment may occur in play.

   4. Intense or prolonged psychological distress at exposure to internal or external cues that symbolize or resemble an aspect of the traumatic event(s).
   5. Marked physiological reactions to internal or external cues that symbolize or resemble an aspect of the traumatic event(s).

C. Persistent avoidance of stimuli associated with the traumatic event(s), beginning after the traumatic event(s) occurred, as evidenced by one or both of the following:
   1. Avoidance of or efforts to avoid distressing memories, thoughts, or feelings about or closely associated with the traumatic event(s).

(*Continued*)

Table 1.1 Diagnostic criteria for posttraumatic stress disorder.—cont'd

2. Avoidance of or efforts to avoid external reminders (people, places, conversations, activities, objects, and situations) that arouse distressing memories, thoughts, or feelings about or closely associated with the traumatic event(s).

D. Negative alterations in cognitions and mood that are associated with the traumatic event(s), beginning or worsening after the traumatic event(s) occurred, as evidenced by two (or more) of the following:
1. Inability to remember an important aspect of the traumatic event(s) (typically due to dissociative amnesia and not to other factors such as head injury, alcohol, or drugs).
2. Persistent and exaggerated negative beliefs or expectations about oneself, others, or the world (e.g., "I am bad," "No one can be trusted," "The world is completely dangerous," "My whole nervous system is permanently ruined").
3. Persistent, distorted cognitions about the cause or consequences of the traumatic event(s) that lead the individual to blame himself/herself or others.
4. Persistent negative emotional state (e.g., fear, horror, anger, guilt, or shame).
5. Markedly diminished interest or participation in significant activities.
6. Feelings of detachment or estrangement from others.
7. Persistent inability to experience positive emotions (e.g., inability to experience happiness, satisfaction, or loving feelings).

E. Marked alterations in arousal and reactivity associated with the traumatic event(s), beginning or worsening after the traumatic event(s) occurred, as evidenced by two (or more) of the following:
1. Irritable behavior and angry outbursts (with little or no provocation) typically expressed as verbal or physical aggression toward people or objects.
2. Reckless or self-destructive behavior.
3. Hypervigilance.
4. Exaggerated startle response.
5. Problems with concentration.
6. Sleep disturbance (e.g., difficulty falling or staying asleep or restless sleep).

F. Duration of the disturbance (Criteria B, C, D, and E) is more than one month.

G. The disturbance causes clinically significant distress or impairment in social, occupational, or other important areas of functioning.

H. The disturbance is not attributable to the physiological effects of a substance (e.g., medication, alcohol) or another medical condition.

In Chapter 3, clinical psychologist Carin Lefkowitz explains how to incorporate AAT into clinical practice. She first discusses the theory and practical aspects of incorporating AAT in clinical practice, then gives case study examples.

In Chapter 4, I discuss PTSD in history, literature, and art. Written records show that PTSD has afflicted our species throughout its history. PTSD is also featured in works of great literature and art. Not surprisingly, Shakespeare has a remarkably complete—as if he got a copy of the DSM centuries in advance—and wonderful description of PTSD in a combat veteran (Hotspur in Henry IV, Part 1). Interestingly, almost all of these early and artistic descriptions of PTSD are in war veterans.

In Chapter 5 James, Whitworth, Ph.D. Associate Professor at the University of Central Florida, a licensed clinical social worker, and a United States Air Force retired Lt. Colonel, discusses the evidence supporting the use of service dogs for military veterans dealing with PTSD. In Chapter 6, Diane Scotland-Coogan, Assistant Professor and clinical social worker at St. Leo University, describes the basis, use, and evidence for equine-assisted therapy for veterans with PTSD.

In Chapter 7, Greg Para, a war veteran and founder of Sarasota Parrot Conservatory, describes his work using avian-assisted therapy to help individuals with PTSD.

In Chapter 8, Simone Emmons bravely tells the story of the rape she suffered while serving in the United States Army during the War on Terror and her subsequent PTSD. She relates how she founded the organization Service Dogs Strong in Maine to provide service dogs for sexual assault survivors with PTSD. Ms. Emmons describes the approach and challenges in treating a sexual assault survivor with PTSD. Ms. Emmons employs and advocates the technique of having the PTSD survivor participate in training their service dog. Trials of this approach are warranted for AAT for PTSD in sexual assault survivors and possibly patients with PTSD due to other causes. Ms. Emmons also highlights the process and challenges of setting up an organization with the mission of providing service animals to individuals with PTSD who experienced sexual assault and trauma and also how common these responses are in this population.

In Chapter 9, Beth Macauley, Ph.D., CCC-SLP, HPCS, Associate Professor in the Department of Communication Sciences and Disorders at Grand Rapids State University, discusses the use of AAT with pediatric patients. In Chapter 10, Sami Abate, Ph.D., MSHS, MSN, RN, CCRN, Director of Research and Nursing Quality at Inspira Health, discusses the use of AAT for patients with heart failure. In Chapter 11, psychiatrist George Grossberg and colleagues discuss the use of AAT in geriatric patients.

In Chapter 12, Jessica Nelson, JD, President and Executive Director of Paws for Heroes, explains their process for selecting and training emotional service canines, and for pairing these dogs with service veterans.

In Chapter 13, Charlotte Glintborg, Ph.D. an Associate Professor in the Center for Developmental and Applied Psychological Science at Aalborg University in Denmark, covers crucial and important ethical issues and issues regarding the welfare of animals participating in ATT. Chapters 7 and 8 on equine-assisted and avian-assisted therapies and also Chapter 10 on AAT for pediatric patients each have sections on animal welfare and the ethical treatment of animals.

In the penultimate chapter (Chapter 14), a United States Army war veteran tells the powerful story of his continued recovery from PTSD due to his companion dog—the dog also having been rescued.

In the final chapter (Chapter 15), we return to Kate's soliloquy about her husband, war veteran Hotspur's PTSD from Shakespeare's Henry IV, Part 1. This chapter uses examples from our war veteran's experience to highlight how AAT can help ameliorate the signs and symptoms of PTSD that Shakespeare so presciently and brilliantly described.

References

Altschuler, E. L. (1999). Pet-facilitated therapy for posttraumatic stress disorder. *Annals of Clinical Psychiatry, 11*(1), 29–30. https://doi.org/10.1023/A:1022808131941

Altschuler, E. (2016). PTSD induced by the trauma of subordinates: The Robert Gates syndrome. *Occupational Medicine, 66*(3), 182. https://doi.org/10.1093/occmed/kqv201

Bisson, J. I., Cosgrove, S., Lewis, C., & Roberts, N. P. (2015). Post-traumatic stress disorder. *BMJ (Online)*, 351. https://doi.org/10.1136/bmj.h6161

Clawson, A. H., Jurbergs, N., Lindwall, J., & Phipps, S. (2013). Concordance of parent proxy report and child self-report of posttraumatic stress in children with cancer and healthy children: Influence of parental posttraumatic stress. *Psycho-Oncology, 22*(11), 2593–2600. https://doi.org/10.1002/pon.3321

Raskind, M. A., Peskind, E. R., Chow, B., Harris, C., Davis-Karim, A., Holmes, H. A., Hart, K. L., McFall, M., Mellman, T. A., Reist, C., Romesser, J., Rosenheck, R., Shih, M. C., Stein, M. B., Swift, R., Gleason, T., Lu, Y., & Huang, G. D. (2018). Trial of prazosin for post-traumatic stress disorder in military veterans. *New England Journal of Medicine, 378*(6), 507–517. https://doi.org/10.1056/NEJMoa1507598

Shalev, A., Liberzon, I., & Marmar, C. (2017). Post-traumatic stress disorder. *New England Journal of Medicine, 376*(25), 2459–2469. https://doi.org/10.1056/NEJMra1612499

CHAPTER 2

History of animal-assisted therapy (AAT) and theories of mechanism of action of AAT for posttraumatic stress disorder (PTSD)

Eric Altschuler[1,2]
[1]Department of Physical Medicine and Rehabilitation, Metropolitan Hospital, New York, NY, United States; [2]Department of Rehabilitation Medicine, New York Medical College, Valhalla, NY, United States

My last year in medical school, I did a rotation at an outpatient Veterans Affairs (VA) clinic. By establishing outpatient clinics such as this, the VA hoped to make it more convenient for veterans to see a doctor by establishing clinics in the community, rather than having them come to their main hospital. One patient I saw had diabetes, hypertension, and other medical problems. When the visit was over, the doctor gave the patient prescriptions for medications to treat his conditions and said that he could pick up the medicines at the pharmacy which was conveniently located across the hallway from the exam room. I left with the patient to take him to the pharmacy, but instead the patient headed for the exit of the clinic. I tried to redirect the patient, but he continued to the exit. I said that it was important for him to pick up his medications and assured him that the cost was covered by his VA benefits. The patient turned to me and said that he wasn't angry, and that he would get his medicines at another pharmacy, but that he was anxious without his dog, and given his posttraumatic stress disorder (PTSD) he was going home.

Was his dog helpful for his PTSD, I asked the patient. Yes, the patient said, his dog helped with the symptoms of his PTSD.

I was surprised, to say the least, to hear the patient say this. At that time, there were not good treatments for PTSD (Shalev, 1997) indeed, we told VA patients with PTSD they needed to "go to their group before they got their medicines." In other words, individual psychotherapy wasn't helpful for PTSD, group therapy wasn't particularly helpful either, and neither were medicines individually or in combination.

Animal Assisted Therapy Use Application by Condition
ISBN 978-0-323-98815-5
https://doi.org/10.1016/B978-0-323-98815-5.00007-0

Animals have surely interacted with humans in health and disease for millennia. One of the first written reports of animals as an adjuvant for medical therapy came from the York Retreat in York, England, in 1792, where pets were used for positive reinforcement for patients to care for themselves (Cusack, 1988). In the 1940s, farm and small animals were incorporated in the treatment plan for soldiers at the Pawling Army Air Force Convalescent Hospital in New York State who were recovering from a variety of medical and psychiatric diseases and conditions (Cusack, 1988). The modern use of animal-assisted therapy (AAT) was galvanized in part by a paper by child psychiatrist Boris Levinson (Levinson, 1962).

I found no published papers or other literature or accounts of dog ownership or AAT as a treatment for PTSD. I published a short note about the patient I had seen at the VA outpatient clinic (Altschuler, 1999). Following my work, a promising preclinical study showing that petting an animal reduced anxiety (Shiloh et al., 2003) Chapter 3 and a proposal of treatment strategy (Lefkowitz et al., 2005) (see also Chapter 3) incorporating AAT into psychotherapy for PTSD were published. AAT for PTSD eventually began to be commonly used for PTSD across the world with anecdotally reported significant benefits (Pandzic, 2012; O'Haire et al., 2015).

In my original paper, I had called for randomized controlled trials (RCTs) of canine-assisted therapy (Altschuler, 1999). In Chapter 5, James Whitworth, LCSW, PhD, an Associate Professor and clinical social worker at the School of Social Work at the University of Central Florida and a retired United States Air Force Lt. Colonel reports on the benefits of canine-assisted therapy for PTSD in war veterans that have been found in increasingly rigorous controlled studies. Canines are not the only animals that are used to treat PTSD. In Chapter 6, Diane Scotland-Coogan, LCSW, PhD from Saint Leo University reports on studies of equine-assisted therapy for veterans with PTSD.

RCTs are most useful and uniquely powerful for assessing the efficacy of a medical intervention while controlling for known and unknown biases. RCTs do not always necessarily compare experimental treatment to an inactive control placebo treatment. Indeed, out of appropriate concern for the risk of suicide in untreated or "placebo-treated" PTSD, the control arm of a recent RCT of canine-assisted therapy for PTSD (U.S. Department of Veterans Affairs, 2020) was an emotional support dog versus a service dog as the treatment arm. Benefit was found, perhaps not surprisingly, in both groups, though more so for subjects in the service dog group. This is still a most useful and important result given the much lower cost of an emotional support dog compared with a service dog. In the case I described of AAT for PTSD (Altschuler, 1999), the patient benefited from "merely" a "regular" pet dog. Due to the very nature of the rigorous requirements for a well-done RCT, information learned from an RCT in terms of interventions, patient groups, and outcome measures can be supplemented by other studies such as case reports or case series. However, doing a case series of even two patients requires institutional review board approval, and the burden of time, paperwork, and cost can be so great as to limit doing these studies.

Given the need for treatments for PTSD (Shalev et al., 2017; Raskind et al., 2018) [1], fortuitously it turns out for AAT for PTSD in combat veterans and others, as I noted (Altschuler, 2018), we can find a tremendous number of cases by simply examining media reports of such. Furthermore, the media reports typically have more detailed information than would ever be allowed to be published in typical medical case reports.

For example, *The New York Post* ran a story (Herbst, 2017) about a US Marine veteran, Bobby Peeker, a combat engineer whose job in Afghanistan was searching for roadside bombs. While deployed, the veteran experienced ambushes, and four men in his company were killed. To try to deal with the PTSD, Peeker started drinking. According to the article, after therapy and medication, he was able to stop drinking, but the PTSD remained. Indeed, the story quotes Peeker as being "in a dark place, a funk." "I was always depressed, constantly thinking about when I was deployed."

Then Peeker was able to obtain a service dog. The dog, Sapper, was extremely helpful in terms of PTSD symptoms and signs. "He's gotten me out of the house," said Peeker. "He's pulled me out of a dark place." And four months after receiving Sapper, Peeker was beginning to be weaned off his medications. Peeker was even able to go back to work in a job repairing roads. But this job would not let him bring Sapper with him. Without

[1] 3, 4-Methylenedioxymethamphetamine (MDMA), the active ingredient in the street drug ecstasy, has been proposed for use to facilitate psychotherapy for the treatment of PTSD and found of benefit in small short-term study (Mitchell et al., 2021). However, along with the logistical demands and costs of the accompanying psychotherapy (Halvorsen et al., 2021) over 40 h and the need for more and long-term efficacy and safety studies, and blinding in such studies (Burke & Blumberger, 2021), there are also concerns (Schenk & Newcombe, 2018) of toxicity, side effects, and abuse potential from the use of MDMA warranting concern and necessitating further study.

Sapper, Peeker had panic attacks so severe it took "lengthy phone calls with his therapist to calm him down," and so Peeker had to leave this job.

Peeker then interviewed for a job with a security company working with a dog to sniff out explosives. This employer agreed to allow Sapper to be trained as a bomb-sniffing dog. The president of the security company described Peeker's and Sapper's job performance, thus "They are excelling one of the best teams we have." This article described the case of an individual with a classic risk factor for developing PTSD personal experience of the trauma of war who developed severe PTSD with severe social and vocational dysfunction and anxiety not relieved by medications or counseling. AAT greatly alleviated his PTSD symptoms, improved his social functioning, and the patient was able to go back to work full time! This case is also a demonstration that a PTSD service animal can be trained to function simultaneously at a high level for other tasks.

Media reports of animals useful in treating PTSD symptoms are not restricted to mammals such as dogs and horses (Kulkarni, 2021), or to print or written journalism: For example, a TV news story (Rogers, 2017) described the case of a combat veteran with PTSD whose symptoms were significantly alleviated by working with parrots. Interestingly, these parrots are at a bird sanctuary after they themselves were rescued after they were traumatized. The veteran in the news story describes how after returning home after his deployment he was too anxious to go outside if he thought he was going to be in a crowd, and he could not drive because cars that might cut into his lane reminded him of explosive-laden cars that had cut into combat convoys. He found working with the parrots alleviated his anxiety and allowed him to do activities he couldn't do previous to working with the birds, and that also his nightmares occurred less often. In Chapter 7, Greg Para, himself a combat veteran and founder of Sarasota Parrot Conservatory, discusses the work he does using avian-assisted therapy to help treat individuals with PTSD.

Finally, returning to canine-assisted therapy, an extremely powerful video (Paws for Veterans, 2014) shows an interview with a US Marine combat veteran with PTSD. The veteran describes how he had severe migraines and anxiety to the point of beginning to load his service revolver to commit suicide. At this point, his dog jumped onto his lap. The veteran pushed the dog away, but the dog would not be deterred. The veteran changed his mind about a suicide attempt. The veteran said that the dog would wake him up in the middle of the night when he was having a nightmare and that he would, initially, only leave the house to go for training with his dog. The veteran's PTSD symptoms have continued to improve, and he credits his dog fully for this.

These media stories are often as detailed sometimes even more detailed than medical case reports. Further, as these are public and often freely and widely distributed, these receive "postpublication" review. Media case reports certainly have the potential problem of selection bias—pet therapy not working out does not sell newspapers or TV advertisements. But there is a similar bias in published medical case reports the file drawer effect of ineffective cases not being submitted for publication. These stories suggest a number of hypotheses about AAT for PTSD that are worthy of future formal study and RCTs: (1) AAT can be dramatically effective in improving PTSD symptoms; (2) there is the potential for benefit from AAT by multiple different animals for PTSD; (3) animals providing AAT may also be dual-purpose animals as Bobby Peeker's service dog Sapper is also a bomb-sniffing dog (Herbst, 2017); (4) AAT may have a role in preventing suicide in patients with PTSD.

How might AAT work for PTSD? Greg Para says that to help treat his PTSD (see Chapter 7), "when working with a parrot, there is a therapeutic value to having a routine with your parrot. So, my healing began with a routine every day. I had to get up to feed Bella, then she would always want to interact with me, and during the day, she was always available; no questions were asked when I needed to be with her. We also had our evening routine where just before dark I would get her and hold her on my chest while we sat on the couch."

Simone Emmons herself an army veteran and sexual trauma survivor says (see Chapter 8) that a service dog helps by constantly being by the side of a PTSD survivor and furthermore waiting and watching for any disturbance or change in the survivor's demeanor.

In his narrative (Chapter 14), the war veteran with PTSD says that his emotional support canine by the name of "Bear" helps him in two ways: (1) Helps in dealing with nightmares: "One of the biggest things Bear has done for me is orient me when I have night terrors. It happens most frequently after I have watched military war movies. I would wake up in the middle of the night and I don't recognize my wife and I am dreaming I am back in Iraq. Bear would jump on the bed and stand between me and my wife. Even though my wife was talking to me, I wouldn't be able to recognize her. It takes a few seconds, but eventually after seeing Bear, I would realize I was home." (2) "I've had Bear for six years. A lot of people wonder what do companion dogs do for veterans like me. To be honest, the main thing with Bear is that he is really not asking anything from you, so it's not an additional load on your plate. He literally just wants to be with you. When your family members might be angry at you because you won't do something or you are upset with them (especially with teenagers!), you have this happy dog that just wants to be petted and shows you pure happiness."

Consistent with these first-person observations, the suggestion (Whitworth et al., 2020) that a service dog provides a bridge for veterans with PTSD to reforming relationships with intimate partners or close friends, and also explaining the presence of PTSD not only in individuals who themselves have experienced trauma (Bisson et al., 2015) but also in bystanders (Erichsen, 1867), caregivers (Clawson et al., 2013), and superiors with PTSD induced by trauma to their subordinates (Gates, 2014; Altschuler, 2016), a theory for the efficacy of AAT for PTSD (Altschuler in (Bahrenburg, 2018)) may be found with reference to a theory of consciousness developed by my teacher in medical school physician and neuroscientist Professor V.S. Ramachandran, MD, PhD (Ramachandran, 2011). Ramachandran notes seven elements of human consciousness: unity, continuity, embodiment, privacy, social embedding, free will, and self-awareness. Ramachandran explains that by "continuity," he means humans perceive oneself to have "a sense of continuity of [their] identity through [their life] time."

A traumatic event breaks this continuity. A caregiving animal may thus help a human suffering from PTSD by establishing a new continuity. Future research on the mechanism of action of AAT in PTSD may provide deep insights into the human condition.

References

Altschuler, E. L. (1999). Pet-facilitated therapy for posttraumatic stress disorder. *Annals of Clinical Psychiatry, 11*(1), 29–30. https://doi.org/10.1023/A:1022808131941

Altschuler, E. (2016). PTSD induced by the trauma of subordinates: The Robert Gates syndrome. *Occupational Medicine, 66*(3), 182. https://doi.org/10.1093/occmed/kqv201

Altschuler, E. L. (2018). Animal-assisted therapy for post-traumatic stress disorder: Lessons from "case reports" in media stories. *Military Medicine, 183*(1–2), 11–13. https://doi.org/10.1093/milmed/usx073

Bahrenburg, C. (2018). Improving therapeutic options for patients with PTSD. *MD Magazine Neurology, 8*(4). www.hcplive.com/view/improving-therapeutic-options-for-patients-with-ptsd.

Bisson, J. I., Cosgrove, S., Lewis, C., & Roberts, N. P. (2015). Post-traumatic stress disorder. *BMJ, 351*, h6161. https://doi.org/10.1136/bmj.h6161

Burke, M. J., & Blumberger, D. M. (2021). Caution at psychiatry's psychedelic frontier. *Nature Medicine, 27*, 1687–1688. https://doi.org/10.1038/s41591-021-01524-1

Clawson, A. H., Jurbergs, N., Lindwall, J., & Phipps, S. (2013). Concordance of parent proxy report and child self-report of posttraumatic stress in children with cancer and healthy children: Influence of parental posttraumatic stress. *Psycho-Oncology, 22*(11), 2593–2600. https://doi.org/10.1002/pon.3321

Cusack, O. (1988). *Pets and mental health*. Haworth Press.

Erichsen, J. E. (1867). *On railway and other injuries of the nervous system*. Henry C. Lea.

Gates, R. (2014). *Duty*. Vintage.

Halvorsen, J.Ø., et al. (2021). Challenges with benchmarking of MDMA-assisted psychotherapy. *Nature Medicine, 27*(10), 1689–1690. https://doi.org/10.1038/s41591-021-01525-0

Herbst, D. (2017). *Bomb-sniffing service dog helps veteran recover from PTSD*. New York Post https://nypost.com/2017/03/06/bomb-sniffing-service-dog-helps-veteran-recover-from-ptsd/ (Original work published 2017).

Kulkarni, K. (July 9, 2021). *DTOM/22 Veterans ranch provides healing through equine therapy, friendship and family*. www.thepublicopinion.com/story/news/2021/07/09/warner-south-dakota-veterans-ranch-healing-equine-therapy-friendship-family-ptsd-autism-dementia/7892975002/.

Lefkowitz, C., Prout, M., Bleiberg, J., Paharia, I., & Debiak, D. (2005). Animal-assisted prolonged exposure. *Society & Animals, 13*, 275–296.

Levinson, B. M. (1962). The dog as a co-therapist. *Mental Hygiene, 46*, 59–65.

Mitchell, J. M., et al. (2021). MDMA-assisted therapy for severe PTSD: a randomized, double-blind, placebo-controlled phase 3 study. *Nature Medicine, 27*(6), 1025–1033. https://doi.org/10.1038/s41591-021-01336-3

O'Haire, M. E., Guérin, N. A., & Kirkham, A. C. (2015). Animal-assisted intervention for trauma: A systematic literature review. *Frontiers in Psychology, 6*. https://doi.org/10.3389/fpsyg.2015.01121

Pandzic, I. (2012). *Animal-assisted therapy and PTSD*. https://archive.org/details/white-paper-animal-assisted-therapy-and-ptsd.

Paws for Veterans. (2014). www.youtube.com/watch?v=PAJAd1r3MAg.

Ramachandran, V. S. (2011). *The tell-tale mrain: a reuroscientist's quest for what makes is human*. W. W. Norton & Co.

Raskind, M. A., Peskind, E. R., Chow, B., Harris, C., Davis-Karim, A., Holmes, H. A., Hart, K. L., McFall, M., Mellman, T. A., Reist, C., Romesser, J., Rosenheck, R., Shih, M. C., Stein, M. B., Swift, R., Gleason, T., Lu, Y., & Huang, G. D. (2018). Trial of prazosin for post-traumatic stress disorder in military veterans. *New England Journal of Medicine, 378*(6), 507–517. https://doi.org/10.1056/NEJMoa1507598

Rogers, J. (January 19, 2017).). *Unique Manatee Co. PTSD treatment program uses parrots*. Channel 8 http://wfla.com/2017/01/19/unique-manatee-co-ptsd-treatment-program-uses-parrots/.

Schenk, S., & Newcombe, D. (2018). Methylenedioxymethamphetamine (MDMA) in psychiatry: pros, cons, and suggestions. *Journal of Clinical Psychopharmacology, 38*, 632–638. https://doi.org/10.1097/JCP.0000000000000962

Shalev, A. Y. (1997). Treatment failure in acute PTSD. *Annals of the New York Academy of Sciences, 821*, 372–387. https://doi.org/10.1111/j.1749-6632.1997.tb48293.x

Shalev, A., Liberzon, I., & Marmar, C. (2017). Post-traumatic stress disorder. *New England Journal of Medicine, 376*(25), 2459–2469. https://doi.org/10.1056/NEJMra1612499

Shiloh, S., Sorek†, G., & Terkel, J. (2003). Reduction of state-anxiety by petting animals in a controlled laboratory experiment. *Anxiety, Stress & Coping, 16*(4), 387–395. https://doi.org/10.1080/1061580031000091582

U.S. Department of Veterans Affairs. (2020). *A randomized trial of differential effectiveness of service dog pairing versus emotional support dog pairing to Improve Quality of Life for Veterans with PTSD.* www.research.va.gov/REPORT-Study-of-Costs-and-Benefits-Associated-with-the-Use-of-Service-Dogs-Monograph1.pdf.

Whitworth, J., O'Brien, C., Wharton, T., & Scotland-Coogan, D. (2020). Understanding partner perceptions of a service dog training program for veterans with PTSD: Building a bridge to trauma resiliency. *Social Work in Mental Health, 18*(6), 604–622. https://doi.org/10.1080/15332985.2020.1806181

CHAPTER 3

Incorporating animal-assisted therapy into treatment for posttraumatic stress disorder

Carin M. Lefkowitz
Private Practice, Philadelphia, PA, United States

Introduction

The notion of incorporating animal-assisted therapy (AAT) into trauma recovery is not a new one. In fact, the first practice of AAT in the United States is recorded as having taken place in 1944 at the Pawling Army Air Force Convalescent Hospital, where war veterans recovered from physical and emotional war wounds (Cusack, 1988). Clinicians have been discussing the merits of AAT for survivors of trauma for decades, and research is beginning to demonstrate what many of us have known intuitively: pleasant interactions with animals can improve functioning in and facilitate treatment for survivors of trauma.

In many ways, research on utilizing AAT for posttraumatic stress disorder (PTSD) and other reactions to trauma is in its infancy. Most of the existing research has focused on treatment for adults (most often military veterans) that incorporates dogs or horses. The methodological rigor of these studies is admittedly low, which limits the conclusions that can be drawn (Fisher et al., 2021; O'Haire et al., 2015; Vitte et al., 2021). Nonetheless, the data suggest that AAT for this population is feasible, effective, and appreciated by clients.

The scientist–practitioner framework enables us to interpret and apply the research to practice based on our clinical knowledge and experience. From that perspective, there exist common-sense strategies to harnessing the benefits of AAT without compromising or conflicting with existing gold standard treatments for trauma recovery. Accordingly, the goals of this chapter are to articulate the clinical rationale for incorporating AAT into psychotherapy and to provide case examples demonstrating its application with adult clients. Although focus is on the treatment of PTSD, the potential benefits of AAT apply to other facets of trauma recovery as well.

Animal Assisted Therapy Use Application by Condition
ISBN 978-0-323-98815-5
https://doi.org/10.1016/B978-0-323-98815-5.20002-5

The benefits of animal-assisted therapy for survivors of trauma

The benefits of positive interactions with animals are innumerable and can contribute to overall well-being. But there are physiological, emotional, cognitive, and social benefits that may be especially helpful to survivors of trauma. Furthermore, these benefits can be harnessed to enhance treatment.

Pleasant interactions with animals have documented physiological benefits for humans. For example, during pleasant interactions with dogs, humans experience an increase in oxytocin and dopamine, which have positive effects on social bonding and mood. Humans can also experience a decrease in cortisol, blood pressure, and heart rate, all of which are associated with a decrease in stress response (Vitte et al., 2021). In fact, there is some evidence that humans' autonomic nervous system stress responses decrease more in the presence of a familiar animal than they do in the presence of a supportive friend or when sitting alone and relaxing (Friedmann et al., 2000; Serpell, 2000).

The emotional benefits of pleasant interactions with dogs go beyond those noted above. An evolutionary theory—the Biophilia Hypothesis—posits that humans have an innate interest in animals and plants due to their evolutionary interdependence. Humans learned how to assess the safety of an environment at least partly by the reactions of nonhuman animals nearby. An animal that seems alert or scared communicates that potential danger is lurking and can trigger a stress response in the human. On the other hand, a calm and relaxed animal conveys a sense of safety and can lead to a relaxation response in the human (Kruger et al., 2004). Importantly, humans appear to not only be aware of animals' stress-relieving benefits but also to actively seek them out. There is evidence that humans will focus on animals more in high-stress situations than in low-stress situations, actively (if not always consciously) seeking the comfort that animals can offer (Friedmann et al., 2000).

Cognitively, positive interactions with dogs provide important information that can influence the expectations, assumptions, and conclusions that a trauma survivor has drawn about themselves, others, and the world. The unconditional positive regard offered by a dog can change the way a survivor views themselves. In a cognitive model, a dog responding to the survivor as someone who is worthy of affection and trust introduces information that is incompatible with negative beliefs that the survivor may hold about themselves, such as being untrustworthy, unlovable, or incapable of connection. Similarly, a survivor developing the ability to read and respond to a dog's behavior, including some that is unpredictable, may foster a sense of competence. This new information or evidence can be utilized in therapy to challenge and restructure unrealistic negative cognitions.

Given that social discomfort and isolation are frequently experienced by individuals with PTSD, the social benefits of animal companionship are extremely relevant. An important factor is that few social skills are required to interact positively with pets. For this reason, a person who struggles with human interactions may find it easier to form a connection with a pet. In turn, receiving positive attention and affection from an animal can meet some of the human's needs for social connection. Pets can also serve as social lubricants by increasing their owner's interactions with other humans. Dog owners may find themselves in social situations more frequently, such as in a park, due to the dog's physiological and social needs. Additionally, social skills developed between a human and pet may ultimately generalize to the human's relationships with others. For example, Abraham et al. (2021) posit that learning how to communicate and demonstrate patience with an animal may be skills that can then be applied to and improve relationships with other humans.

Considering these benefits, it is easy to extrapolate how animals may contribute to positive treatment experiences. Applying the Biophilia Hypothesis, we can imagine the effect that a calm animal in a therapy office has on a client; the animal signals that the office is a safe place, and the therapist is a safe person (Kruger et al., 2004). It is consequently unsurprising that therapists utilizing AAT often appear less threatening and more empathic to clients (Fine, 2000) and that AAT may facilitate a faster development of rapport between client and therapist (Jenkins, 2001). AAT may further facilitate engagement in treatment by appealing to clients who are wary of traditional psychotherapies and who prefer alternative care options. The physiological, emotional, and cognitive benefits can be capitalized upon in therapy for PTSD, the most effective of which aim to reduce physiological arousal, process emotions, and restructure overly negative cognitions (International Society for Traumatic Stress Studies Guidelines Committee, 2019; Management of Posttraumatic Stress Disorder Working Group, 2017).

Perhaps, one of the most valuable impacts of AAT may be on attrition from psychotherapy. Research demonstrates that several highly effective psychotherapies exist for PTSD. Unfortunately, the research also demonstrates that patients may drop out of trauma-focused treatment prematurely, with an average 36% attrition rate (Imel et al., 2013). And that says nothing of the proportion of clients who avoid trauma-focused interventions altogether. While many explanations have been offered—from competing life demands to the challenging nature of therapy itself—the result is that many clients are not receiving the best possible treatment. AAT has the potential to improve this reality. Several studies highlight the ways in which animals bolster clients' motivation for treatment. In one study of a service dog training intervention (Abraham et al., 2021), one participant said, "every day I look forward to coming here and I want to continue doing this." Every participant in the program reported that they either would or already had recommended the program to others. In a study of a group intervention, Beck, Seraydarian, and Hunter (Beck et al., 1986) found that participants attended sessions more consistently and were more active participants when the group included animal-assisted interventions.

Taken together, AAT offers benefits that are especially well suited to clients recovering from trauma. AAT has relevant benefits for the individual client and can also facilitate engagement in treatment that can be challenging at times.

The state of the research

Research is beginning to provide data to support these theoretical models of AAT for trauma recovery. While a comprehensive review of the literature is outside the scope of this chapter, it is important to understand the overall state of the literature. Several recent studies looked specifically at the impact that equine- and canine-assisted therapy can have on PTSD and other trauma-related responses.

A study of equine-assisted therapy for PTSD (EAT-PTSD) investigated the effectiveness of a weekly group intervention (Fisher et al., 2021). The researchers posit that horses are ideal choices for AAT because they are naturally sensitive and responsive to both verbal and nonverbal cues received from humans. The EAT-PTSD protocol is not trauma-focused. The goals of this experiential treatment are to increase emotional awareness, improve communication, regulate emotions, improve problem-solving skills, and increase self-confidence through prescribed interactions with horses (i.e., leading a horse onto a tarp). Participants (who were all military veterans) reported a 54% improvement in reported PTSD symptoms and had a low (8%) dropout rate. A connected study sought to identify any neurobiological changes that might account for the improvements experienced by participants (Zhu et al., 2021). Utilizing neuroimaging, they found that participants who completed EAT-PTSD demonstrated changes in connectivity in the basal ganglia network, which are involved in reward processing, anhedonia, and learning. The change in connectivity may account for the reported improvement in symptoms.

Studies of canine-assisted therapy for PTSD have predominantly looked at the effects of having a companion dog, or having or being involved in the training of a service dog (Vitte et al., 2021). The outcomes generally demonstrate improvement in symptoms of PTSD (particularly a decrease in anxiety and improved grounding abilities in the presence of intrusion symptoms), social functioning, and quality of life. These studies also boast low attrition rates.

Despite the compelling rationale behind AAT for survivors of trauma and the public interest in this topic, the field is nascent in many ways. While researchers are eager to demonstrate the merits of AAT, published studies lack the methodological rigor to confidently draw conclusions about how the effectiveness of AAT compares to treatment as usual. Fisher and colleagues (Fisher et al., 2021) state, "Extant EAT research focusing on … PTSD … is scarce and generally poorly designed, characterized by small sample sizes, inconsistent assessments, unstandardized treatment procedures, and researcher conflicts of interest." Similarly, in a recent systematic review of canine-assisted treatment for PTSD, the authors could not draw strong conclusions, given the variability and limitations within the studies (Vitte et al., 2021). The existing literature, including the studies noted in this chapter, consists largely of anecdotal and subjective self-reports. Few studies were peer-reviewed or employed randomized control procedures (O'Haire et al., 2015). While multiple studies (but not all) demonstrated promising results when utilizing dogs as a complementary treatment to standard therapy, there were high risks of bias in most studies and the interventions were not described well enough to allow for replication. Also, differing methodologies made it difficult to compare studies. Importantly, most studies focused on military veteran populations, which also limits generalizability.

I think it fair to conclude that while the emerging literature on AAT for PTSD is promising, the existing data does not provide definitive guidelines for how to incorporate AAT in treatment at this time. The recommendations I offer here are based on the existing research combined with my knowledge and experience with evidence-based psychotherapies for PTSD and my theoretical understanding of how AAT can be most usefully integrated with them.

A brief summary of evidence-based psychotherapies for PTSD

The literature on effective treatments for PTSD is quite mature. Decades' worth of high-quality research has produced several efficacious treatment options for clients. Several major institutions (American Psychological Association, 2017; International Society for Traumatic Stress Studies Guidelines Committee, 2019; Management of Posttraumatic Stress Disorder Working Group, 2017) have published clinical practice guidelines that analyze the quality and quantity of data and offer best practice recommendations. Cognitive-behavioral trauma-focused treatments are consistently recommended as first-line interventions for PTSD. These include (but are not limited to) cognitive processing therapy (CPT), prolonged exposure (PE), and eye movement desensitization and reprocessing (EMDR).

CPT (Resick et al., 2017) addresses PTSD from a predominantly cognitive framework, helping clients to articulate and restructure rigid and unrealistic beliefs. Identifying and processing primary emotions like sadness and grief is also a proposed mechanism of action in CPT. Most of the interventions in CPT are introduced via worksheets or other written assignments, each of which build upon each other throughout treatment. Other treatments such as PE (Foa et al., 2019) and EMDR (Shapiro, 2017) have a more behavioral focus with the practice of exposure. Exposure to and engagement with painful emotions and avoided memories and stimuli lead to habituation and emotional processing. Cognitive restructuring aims to reinterpret unhelpful beliefs that are elicited during the sessions and at-home practice.

While each of these treatments target PTSD in slightly different ways, they all have several characteristics in common:

1. They are manualized treatment protocols that are designed to be delivered in a step-by-step fashion.
2. They are trauma-focused and therefore include some explicit discussion of the traumatic event.
3. They involve emotional engagement with the trauma memory.
4. They involve restructuring of unrealistic or unhelpful cognitions.
5. A strong therapeutic relationship is crucial.

Although these established evidence-based practices (EBPs) should be the first-line recommendation for clients with PTSD, the reality is that a percentage of clients do not respond to treatment, some discontinue treatment early, and another subset are reluctant to engage altogether. Therefore, it is reasonable to explore adjunctive treatments that may facilitate EBPs or offer relief for patients who refuse to engage in EBPs. Given the benefits described previously, AAT is uniquely suited to complement common EBP interventions, expedite the development of a strong therapeutic relationship, and facilitate engagement from clients who may otherwise be reluctant.

Models for incorporating AAT into evidence-based psychotherapies for PTSD

The existing literature describes how dogs have served therapeutic roles in countless ways. At the most unstructured end of the spectrum, the therapist's dog may simply be present in a therapy session, providing visual and tactile comfort and reassurance to a client. A client's own dog may provide personal emotional support, behavioral activation, and opportunities for social engagement. At the most structured extreme, a service dog is trained to perform specific tasks relevant to the client's health and well-being. Finally, a dog may be utilized as part of a structured psychotherapy program that incorporates the dog in the application of therapeutic tasks and skills.

While the role that a dog plays in treatment may vary, the safety of everyone involved is paramount. Some guidelines and precautions are therefore universal. Canine-assisted interventions should only take place with a well-trained dog whose health is routinely monitored by a veterinarian. While there are no definitive obedience guidelines for dogs who work in AAT (except for service dogs), they should minimally be trained in basic obedience skills such as walking on a leash, sitting, and being housebroken. The dog's individual temperament should also be considered. Dogs who actively participate in therapy sessions should have all vaccinations kept up to date to prevent the spread of zoonotic diseases. They should also be provided a "safe space" within the therapy office that they can retreat to if they feel overwhelmed. This greatly reduces the likelihood of bites and scratches, as dogs are more likely to do so if they feel threatened without an option for escape. Obviously, canine-assisted therapy is most appropriate for clients who have positive, or at least neutral, associations with dogs. The client's attitudes toward dogs should be assessed prior to any canine-assisted activity.

Unstructured AAT may occur in a treatment session either with the therapist's trained dog or the client's pet dog. The goal of AAT in this scenario is that the dog provides comfort while the client engages in treatment for their PTSD. The mere presence of a calm dog can reduce physiological and emotional arousal within the session. Physical contact with the dog, through petting or sitting with the client, can further provide tactile comfort. In this setting, the presence of the dog potentially facilitates the client and therapist's preferred treatment approach by signaling that the therapy office is a safe place, reducing emotional and physiological distress, and reducing attrition.

As noted earlier, first-line treatments for PTSD are typically trauma-focused and incorporate cognitive and/or behavioral techniques (International Society for Traumatic Stress Studies Guidelines Committee, 2019; Management of Posttraumatic Stress Disorder Working Group, 2017). AAT can complement these EBPs in many ways. A recent study (Wharton et al., 2019) piloted an adaptation of CPT that incorporates equine-facilitated interventions: equine-facilitated cognitive processing therapy (EF-CPT). The researchers theorized that equine-facilitated therapy provides mechanisms for increased physical activity and externalization of negative affect, as well as opportunities to experientially address challenges common to individuals with PTSD, including frustration tolerance and trust building. Twenty-seven veterans with PTSD received the manualized treatment that combined the standard 12-session CPT protocol with suggested equine-facilitated activities. These activities were designed to mirror the skills taught during the accompanying CPT session. For example, session 1 of CPT involves psychoeducation regarding the fight/flight/freeze response and discussion of how it has manifested in the client's life. One of the suggested equine activities for this session includes discussion of the fight/flight reaction in horses. This discourse has the power not only to normalize the fight/flight response as a universal response across species but also to help the client articulate advanced logical thinking resources that he has to address emotional dysregulation. Other equine-facilitated activities offer opportunities for the client to apply and practice concepts and skills learned in CPT, such as setting flexible boundaries and exerting power and control in healthy ways (Wharton et al., 2015).

The results from this pilot study demonstrate that EF-CPT was effective in reducing self-reported symptoms of PTSD. At the end of the study, 84% of participants no longer met the criteria for PTSD. Additionally, there was a very small attrition rate (7%) from the intervention. Because the design of this study did not include randomization or a control condition, conclusions cannot be definitively drawn about how its effectiveness and attrition compare to the standard CPT protocol. Nonetheless, this study demonstrates the feasibility of formally combining AAT with an EBP for PTSD and provides an exciting model for future study.

While a similar model does not exist for canine-assisted CPT, there are several ways that a dog can serve as an adjunct to CPT. The unstructured interventions noted earlier could apply to CPT sessions. A dog can offer comfort while a patient processes natural emotions or discusses painful trauma-related cognitions. The calming presence of the dog may also help patients engage in the more challenging tasks of CPT, such as reading their impact statement or trauma account aloud. Planned activities with the dog may also be "assigned" as part of the "Esteem Module," wherein the client is asked to engage in enjoyable and rewarding activities daily.

Dogs can also assist in exposure-based interventions. In the case of PE (Foa et al., 2019), one form of exposure work is engaging in in vivo (real-world) situations that the client has previously avoided or tolerated with great distress. Some common examples include visiting a crowded public space or watching a movie that involves violence. The therapist works with each client to collaboratively develop an individualized in vivo hierarchy that identifies avoided situations and the client's anticipated distress in each situation. Over the course of PE, the client will consistently and repeatedly engage in these tasks in a hierarchical fashion, beginning with moderately distressing situations and progressing to increasingly difficult tasks. The client's level of distress is measured using the Subjective Unit of Discomfort Scale (SUDS). The client rates anticipated distress on a 0—100 scale, where "0" = no discomfort and "100" = the most distressed the client has ever been. Having a trusted friend or family member accompany the client during in vivo exposure tasks is a way to calibrate how much distress is likely to be experienced. For example, a client may experience significantly less distress when visiting a crowded park if they are accompanied by a trusted friend (SUDS = 45), compared to if the client goes to the park alone (SUDS = 70). Based on the behavioral model, being accompanied by a companion is a "safety behavior" that will ultimately be phased out, so that the client can engage in each task on their own. Although the PE manual specifically refers to human companions, there is no theoretical argument against allowing the companion to be a dog. While building the in vivo hierarchy, the client and therapist can discuss how the presence of the dog changes the client's anticipated level of distress and specifically articulate how and when the dog will accompany the client (i.e., sitting with the client on a bench in the middle of the park). The obvious limitations are based on safety concerns (Is this a safe situation for the dog?) and relevant laws (dogs who are not designated service dogs may not be permitted in indoor public spaces, depending on jurisdiction). As with a human companion, the presence of the dog should be predictably phased out over time.

Dogs can also offer support during imaginal exposure, both in and outside of sessions. Imaginal exposure is the process by which the client vividly recalls their index trauma, sharing details, emotions, and thoughts in the present tense. Imaginal exposure is practiced in each session starting at session #3 and is followed by cognitive restructuring aimed at maladaptive thoughts that arise during and after the client revisits their trauma memory. PE sessions are audio recorded, and the client is asked to listen to the recording of the imaginal exposure daily in between sessions. This repeated exposure facilitates emotional habituation. Ideally, the client is fully engaged with the memory during imaginal exposure, so tactile or visual contact with a dog during this exercise may be more distracting than helpful. A dog can, however, provide comfort and support during the cognitive restructuring portion of the session and/or recording. In the case of a truly reluctant client, the benefits of temporarily having a dog present during imaginal exposure may outweigh the risks. However, a clear plan for phasing out the dog should be articulated from the outset, as a therapist would do with any safety behavior.

It is important that the role of the dog be discussed openly regardless of which treatment approach is utilized. True informed consent includes discussion of the client's preferences regarding AAT, availability of other treatment options, and a research-informed assessment of potential risks and benefits. A legitimate risk in AAT is that a client will attribute success of the treatment to the presence of the dog, rather than to the hard work the client has engaged in. This may be especially true of clients who tend to minimize their own skills and accomplishments. The danger is that a patient will end treatment with the new belief "I can do difficult things, but only with a dog." Or "painful emotions are only tolerable when I have my dog." In a behavioral treatment model, this outcome can be prevented with the planned phasing out of the presence of the dog during treatment sessions and at-home practice. In all treatment models, the therapist should remain vigilant for signs that the client is externalizing accountability for progress. Cognitive therapy techniques can be helpful in addressing misattribution of progress, and praise and encouragement for the client's efforts should be offered consistently.

Case examples

Over the years, I have had the pleasure of incorporating canine-assisted interventions into my practice, both formally and informally. I share the following two case examples with the goal of illustrating theoretically sound methods of incorporating AAT into the treatment of PTSD.

Phil and Fonzie: the dog that aided exposure therapy

I met "Phil" when he was 62 years old. A military veteran, he had been living with combat-related PTSD for nearly 40 years. He was service connected (receiving disability through the Veterans Affairs) for PTSD but had never sought treatment before. He sought out psychotherapy after a fellow veteran told Phil that treatment had worked for him. At our intake appointment, Phil reported intrusive memories of a lethal vehicle accident, nightmares, startle response, hypervigilance, avoidance of reminders of the war, depressed mood, and irritability. After his intake and a few psychoeducation sessions, Phil became interested in PE. Although he was eager for treatment and understood the rationale, he had reasonable apprehension about the imaginal and in vivo exposure tasks. In preparing for the challenging work ahead, we brainstormed about what resources he had to cope with any temporary increase in symptoms that might occur. That is when Phil talked about his terrier, Fonzie. Fonzie seemed attuned to Phil's emotional state and would sit on his lap and lick Phil's face if he sensed distress. Phil described Fonzie as an emotional comfort and noted that his anxiety and distress decreased in Fonzie's presence. Phil felt he would be more able and willing to approach the memories and stimuli he had been avoiding if Fonzie accompanied him, at least initially. We therefore incorporated Fonzie into PE as a companion. We identified specific tasks that Fonzie could assist with, as well as a timeline for phasing out his presence. For example, when listening to audio recordings of our sessions for homework, Phil would have Fonzie sit on his lap the first time he listened to the recording each week. Having overcome this first hurdle, Phil would then listen to the recordings without Fonzie in the room for the rest of the week. Fonzie also accompanied Phil during some in vivo exposure assignments. One item on Phil's exposure hierarchy was sitting near an open flame, such as next to a fireplace or burning candle, which reminded Phil of the vehicle accident and subsequent fire. Having added "sitting in a room with a burning candle" onto his hierarchy, we established SUDS ratings based on whether Fonzie was on his lap (45) or not (80). This task was then added onto his hierarchy as two different items that Phil would engage in over the course of PE. Another option would have been to approach this task in the same way he did his imaginal exposure homework: to have Fonzie on his lap for the first practice of sitting in a room with a lit candle and then sitting without Fonzie in subsequent exposures.

Phil successfully completed PE and experienced a significant decrease in his PTSD symptoms, as well as comorbid depressive symptoms. He reported improved functioning in social and family spheres. While he accepted credit for choosing to engage in the treatment and completing the challenging work, he consistently noted that Fonzie's presence increased his willingness. In fact, at the conclusion of PE, Phil enrolled Fonzie in obedience training, so that he could safely serve as an emotional support dog for other veterans. Once Fonzie completed his training, Phil acquired an "emotional support dog" badge for Fonzie and began bringing him to appointments at the clinic, as well as to other medical appointments. Fonzie was an immediate hit. On days that Phil brought Fonzie to the clinic, the dyad would quickly become surrounded in the waiting room by other clients who wanted to pet Fonzie or talk about him. Consequently, Fonzie's contributions extended well beyond the PE protocol. Phil increased his social engagement with others as he answered questions about Fonzie in multiple settings. And the social benefits extended to the patients in the waiting rooms, who often felt safe engaging with Phil and Fonzie despite their typical social discomfort.

Ben and clover: the dog that offered general support

Despite my preference for treating PTSD with EBPs, this is not always an option. Perhaps the client is unwilling to engage in an EBP or has more pressing concerns that need to be addressed first. Other times I may meet with a client who has already completed an EBP but has residual needs. AAT can be utilized as an adjunct to other treatment modalities when EBPs are not an appropriate option, and it can support recovery in other ways. My work with "Ben" illustrates this approach.

I met Ben, a 30-year-old man, when he had already been suffering with PTSD for about two years. Ben was on disability from his job in law enforcement. He sought treatment following a physical altercation with a suspect during which he sustained a concussion and other debilitating physical injuries. I began working with him about a year and a half after the incident, at which point he was still in significant pain despite multiple surgeries and physical therapy. At his intake, he reported symptoms consistent with PTSD, including intrusive memories, nightmares, insomnia, avoidance of crowded locations and violent movies, startle response, depressed mood, and irritability. Ben completed a course of CPT with me and made great progress in his negative cognitions and depressed mood. Unfortunately, he had ongoing legal and medical stressors related to his index trauma, so it was never fully in his "rearview mirror." He continued to cope with symptoms of PTSD at the conclusion of CPT, albeit decreased in severity and frequency. Throughout treatment, Ben talked about his husky, Clover, and how her presence comforted him in times of stress. Unlike PE, CPT includes few behavioral interventions. Therefore, there was not an active role for Clover to play in this type of treatment. But while she did not play a role in our CPT sessions, she provided several social and physical benefits that supported Ben's recovery. Perhaps most notably, Clover presented many (sometimes forced) opportunities for behavioral activation. As a husky, Clover needs a good deal of physical activity and releasing her to run around the backyard is not enough. Oftentimes, the only reason Ben would leave his house (and sometimes even just his bedroom) was to take Clover for a walk. If he felt particularly agitated toward his roommates, he would go for a drive with Clover or take her on a hike. Even after Ben completed CPT, he tried to expand upon his self-imposed limits by getting out of the house more often and visiting public spaces. We set specific goals that included Clover, such as playing with her outside and taking her to dog-friendly stores. On several occasions, he took Clover to busy pet stores and big box home improvement stores. He described feeling more relaxed in this setting, trusting that Clover "had my back." Clover's presence also facilitated social engagement, as other customers want to interact with her. Ben noted, "people always have questions about Clover because she's so pretty." He reported a greater sense of social comfort in these situations; he feels at ease when attention is on Clover instead of him, and her presence eases the tension he feels when socializing with others. In these ways, Clover extended upon the specific goals and interventions that Ben and I worked on during CPT. Ben's relationship with Clover organically sparked social engagement and increased physical activity, goals that are not central to CPT but which are highly beneficial for recovery from PTSD. As Ben continued to set behavioral goals for himself, we found ways for Clover to assist him in achieving them.

Conclusions

A thoughtfully constructed treatment plan that incorporates AAT can offer considerable benefits to survivors of trauma. The presence of a calm and friendly animal can reduce the effects of anxiety, depression, and loneliness in clients while improving physical health, increasing behavioral activation, and providing a sense of companionship. These potential benefits are uniquely suited to survivors of trauma, who often feel unsafe around other humans and are therefore prone to becoming socially isolated.

A theoretically sound model of AAT for PTSD considers how the presence of an animal will facilitate and enhance specific treatment goals and interventions. Additionally, consideration must be given to whether the presence of the animal detracts from treatment goals or interventions, as could be the case with their ongoing presence during exposure assignments. Based on the research that currently exists, AAT is most ethically viewed as an adjunct to evidence-based treatments. That said, high attrition rates and some clients' reluctance to engage in trauma-focused treatments result in limited accessibility and engagement. The adjunctive value of AAT should therefore not be minimized, as it may engage and retain clients who might not otherwise receive a full "dose" of effective treatment.

References

Abraham, P. A., Kazman, J. B., Bonner, J. A., Olmert, M. D., Yount, R. A., & Deuster, P. A. (2021). *Effects of training service dogs on service members with PTSD: A pilot-feasibility randomized study with mixed methods*. Military Psychology. https://doi.org/10.1080/08995605.2021.1984126

American Psychological Association. (2017). *Clinical practice guidelines in the treatment of PTSD in adults*. American Psychological Association. www.apa.org/ptsd-guideline/ptsd.pdf.

Beck, A. M., Seraydarian, L., & Hunter, G. F. (1986). Use of animals in the rehabilitation of psychiatric inpatients. *Psychological Reports, 58*(1), 63–66. https://doi.org/10.2466/pr0.1986.58.1.63

Cusack, O. (1988). *Pets and mental health*.

Fine, A. (2000). Animals and therapists: Incorporating animals in outpatient psychotherapy. In A. Fine (Ed.), *Handbook on animal-assisted therapy: Theoretical foundations and guidelines for practice* (pp. 179–208). US Academic Press.

Fisher, P. W., Lazarov, A., Lowell, A., Arnon, S., Turner, J. B., Bergman, M., Ryba, M., Such, S., Marohasy, C., Zhu, X., Suarez-Jimenez, B., Markowitz, J. C., & Neria, Y. (2021). Equine-assisted therapy for posttraumatic stress disorder among military veterans. *The Journal of Clinical Psychiatry, 82*(5). https://doi.org/10.4088/jcp.21m14005

Foa, E., Hembree, E. A., Rothbaum, B. O., & Rauch, S. (2019). *Prolonged exposure therapy for PTSD: Emotional processing of traumatic experiences—Therapist guide*. Oxford University Press. https://doi.org/10.1093/med-psych/9780190926892.001.0001

Friedmann, Thomas, S. A., & Eddy, T. J. (2000). Companion animals and human health: Physical and cardiovascular influences. In A. L. Podberscek, E. S. Paul, & J. A. Serpell (Eds.), *Companion animals and us: Exploring the relationship between people and pets* (pp. 125–142). Cambridge University.

Imel, Z. E., Laska, K., Jakupcak, M., & Simpson, T. L. (2013). Meta-analysis of dropout in treatments for posttraumatic stress disorder. *Journal of Consulting and Clinical Psychology, 81*(3), 394–404. https://doi.org/10.1037/a0031474

International Society for Traumatic Stress Studies Guidelines Committee. (2019). *Posttraumatic stress disorder prevention and treatment guidelines: Methodology and recommendations.* https://istss.org/getattachment/Treating-Trauma/New-ISTSS-Prevention-and-Treatment-Guidelines/ISTSS_PreventionTreatmentGuidelines_FNL-March-19-2019.pdf.aspx.

Jenkins, Missel (2001). The use of choice theory in animal assisted therapy for children and young adults. *International Journal of Reality Therapy, 20*(2), 40–41.

Kruger, K. A., Trachtenberg, S. W., & Serpell, J. A. (2004). *Can animals help humans heal? Animal-assisted interventions in adolescent mental health.* Philadelphia, PA: The Meeting of the Center for Interaction of Animals and Society (The Center for the Interaction of Animals and Society).

Management of Posttraumatic Stress Disorder Working Group. (2017). *VA/DoD clinical practice guidelines for the management of posttraumatic stress disorder and acute stress reaction.* Version 3.0 www.healthquality.va.gov/guidelines/MH/ptsd/VADoDPTSDCPGFinal012418.pdf.

O'Haire, M. E., Guérin, N. A., & Kirkham, A. C. (2015). Animal-assisted intervention for trauma: A systematic literature review. *Frontiers in Psychology*, 6. https://doi.org/10.3389/fpsyg.2015.01121

Resick, P. A., Monson, C. M., & Chard, K. M. (2017). *Cognitive processing therapy for PTSD: A comprehensive manual.* The Guilford Press.

Serpell, J. (2000). Creatures of the unconscious: Companion animals as mediators. In A. L. Podberscek, E. S. Paul, & J. A. Serpell (Eds.), *Companion animals and us: Exploring the relationships between people and pets* (pp. 109–121). Cambridge University.

Shapiro, F. (2017). *Eye movement desensitization and reprocessing (EMDR) therapy: Basic principles, protocols, and procedures* (3rd ed.). Guilford Press.

Vitte, P., Bragg, K., Graham, D., Davidson, J., Bratten, T., & Angus-Leppan, G. (2021). The role of canines in the treatment of posttraumatic stress disorder: A systematic review. *Psychological Trauma: Theory, Research, Practice, and Policy, 13*(8), 899–906. https://doi.org/10.1037/tra0001074

Wharton, T. C., Scogin, F. R., & Malone, M. (2015). *Equine-facilitated cognitive processing therapy: An addendum to the CPT Patient Workbook.* Florida: Veteran/Military Version (Author).

Wharton, T., Whitworth, J., Macauley, E., & Malone, M. (2019). Pilot testing a manualized equine-facilitated cognitive processing therapy (EF-CPT) intervention for PTSD in veterans. *Psychiatric Rehabilitation Journal, 42*(3), 268–276. https://doi.org/10.1037/prj0000359

Zhu, X., Suarez-Jimenez, B., Zilcha-Mano, S., Lazarov, A., Arnon, S., Lowell, A. L., Bergman, M., Ryba, M., Hamilton, A. J., Hamilton, J. F., Turner, J. B., Markowitz, J. C., Fisher, P. W., & Neria, Y. (2021). Neural changes following equine-assisted therapy for posttraumatic stress disorder: A longitudinal multimodal imaging study. *Human Brain Mapping, 42*(6), 1930–1939. https://doi.org/10.1002/hbm.25360

CHAPTER 4

Posttraumatic stress disorder in history, literature, and art

Eric Altschuler[1,2]
[1]Department of Physical Medicine and Rehabilitation, Metropolitan Hospital, New York, NY, United States; [2]Department of Rehabilitation Medicine, New York Medical College, Valhalla, NY, United States

In Act 2, Scene 3 of Shakespeare's Henry IV, Part 1 (1598), Kate, the wife of Hotspur, gives the monologue below. Henry Percy ("Hotspur") is an important character in the play (Altschuler, 2017) who, with his family, helped King Henry IV wrest the throne from Richard II, and then, in extensive battles, fights on behalf of Henry, successfully defeating a Scottish rebellion. Later Hotspur crosses purposes with King Henry and is eventually defeated in solo combat by Henry's son Hal, the future King Henry V.

O, my good lord, why are you thus alone?
For what offence have I this fortnight been
A banish'd woman from my Harry's bed?
Tell me, sweet lord, what is't that takes from thee
Thy stomach, pleasure and thy golden sleep?
Why dost thou bend thine eyes upon the earth,
And start so often when thou sit'st alone?
Why hast thou lost the fresh blood in thy cheeks;
And given my treasures and my rights of thee
To thick-eyed musing and curst melancholy?
In thy faint slumbers I by thee have watch'd,
And heard thee murmur tales of iron wars;
Speak terms of manage [horsemanship] to thy bounding steed;
Cry 'Courage! to the field!' And thou hast talk'd
Of sallies and retires, of trenches, tents,
Of palisadoes, frontiers, parapets,
Of basilisks, of cannon, culverin,
Of prisoners' ransom and of soldiers slain,
And all the currents of a heady fight.
Thy spirit within thee hath been so at war,
And thus hath so bestirr'd thee in thy sleep,
That beads of sweat have stood upon thy brow

Animal Assisted Therapy Use Application by Condition
ISBN 978-0-323-98815-5
https://doi.org/10.1016/B978-0-323-98815-5.00002-1

Like bubbles in a late-disturbed stream;
And in thy face strange motions have appear'd,
Such as we see when men restrain their breath
On some great sudden hest. O, what portents are these?
Some heavy business hath my lord in hand,
And I must know it, else he loves me not.

The poetry is beautiful, and Kate clearly truly loves Hotspur and is concerned about him. Is there a different perspective that might provide another understanding or insight into this passage? There is. That perspective is an appreciation that this soliloquy represents the first complete description of the disease posttraumatic stress disorder (PTSD) remarkably centuries before its first appearances in medical journals:

In 1915, a French physician reported on four soldiers with "battle hypnosis" (Milian, 1915), and a British physician reported on three soldiers who suffered "loss of memory, vision, smell, and taste" (Myers, 1915) describing the condition as "shell shock." In 1916, a German psychiatrist, described the symptoms of sudden muteness, deafness, and inability to stand or walk (Kaufmann, 1916) in soldiers. All of the above were considered symptoms of "shell shock" and were present in many soldiers who fought in World War I (Salmon, 1917). Similar symptoms were present in soldiers in World War II (Grinker & Spiegel, 1945), though often different names such as "war neuroses" or "operational fatigue" were used to describe them.

In 1952, symptoms such as shell shock or war neuroses were listed in first version of the Diagnostic and Statistical Manual (DSM-I) under the category of "gross stress reaction." This manual which is the primary reference for psychiatric diagnoses is now in its fifth edition as the DSM-5. Strangely, this category was removed from the DSM-II. The experiences of numerous war veterans, including many US Vietnam war veterans, and also the recognition that posttraumatic symptoms can be experienced not only by war veterans but also by individuals who have experienced traumatic events such as a sexual assault or natural disaster (symptoms can also be experienced by loved ones or caregivers of individuals who have experienced a traumatic event (Clawson et al., 2013)), led to the current diagnosis of PTSD, which appeared in the DSM-III (Bentley, 2005; Altschuler, 2017). The DSM-5 criteria for PTSD are listed in Table 1.1.

For his medical school dissertation, Swiss physician Johannes Hofer (1669–1752) described the concept of "nostalgia" (1688; University of Basel). Hofer's definition included melancholy, incessant thinking of home, disturbed sleep or insomnia, weakness, loss of appetite, anxiety, cardiac palpitations, stupor, and fever. Thousands of soldiers in the US Civil War were diagnosed with "nostalgia." In 1871, Jacob Mendez da Costa described a constellation of symptoms in soldiers often occurring together—chest-thumping (tachycardia/palpitations), anxiety, and breathlessness—referred to as "soldier's heart" (Da Costa, 1871).

Another early name for PTSD was "railway spine": a syndrome of posttraumatic mental symptoms in individuals who had survived the then not uncommon railroad accidents (Erichsen, 1867).

The history of the human race is one of war, and PTSD seems to accompany all wars, so we would expect to find descriptions of PTSD throughout our written history (Altschuler, 2017). A problem in finding such evidence is that as time recedes, there is less and less textual evidence and only a fraction of texts survive. Nevertheless, there are some hints of PTSD many centuries ago and in diverse geographic locations—the ancient Near East (Abdul-Hamid & Hughes, 2014) and the Indian subcontinent (Jayatunge, 2010), for example. The clearest description of PTSD from the deep historical past is an account from Mesopotamia more than 4000 years ago in which Ezra noted (Ezra, 2001) that a cuneiform tablet (Fluckiger-Hawker, 1999) described the Sumerian King Urnamma's death by battle (c. 2040 BCE), and the consequence of his death includes a description of PTSD.

From the [... the ...] was [...] evenly in/on the land. [The ...] struck, the palace(s) was collapsed. [The ...] spread panic rapidly among its Black-headed who dwelt there. [The ...] established its abandoned places in Sumer. In its vast [...] cities are destroyed, the people are seized with panic. Evil came upon Ur ... They weep bitter tears in their broad squares where merriment had reigned. With their bliss (fullness) having come to an end, the people do not sleep soundly.

More recently, Jean Froissart (1337?—1400/01)—the most representative chronicler of the Hundred Years' War (1337—1453) between England and France—sojourned in 1388 at the court of Gaston Phoebus, Comte de Foix. He described the case of the Comte's brother, Pierre dc Beam, who could not sleep near his wife and children because of his habit of getting up at night and seizing a sword to fight oneiric enemies (Croq & Croq, 2000).

Great writers were aware of the disease as well, and literary "cases" also go back in time. For example, Huber and Te Wildt make an interesting case that Dr. Manette from Dickens' A Tale of Two Cities experiences dissociative traumatic reexperiences in the context of PTSD (Huber & Te Wildt, 2005).

Let's look at Shakespeare's description again with side notes referring to the DSM criteria (Table 1.1) and features of PTSD. (Annotations adapted from those of clinical psychiatrist Jonathan Shay (Shay, 1994)):

O, my good lord, why are you thus alone? [social withdrawal]
For what offence have I this fortnight been [random rage at family]
A banish'd woman from my Harry's bed? [sexual disinterest/disfunction]
Tell me, sweet lord, what is't that takes from thee
Thy stomach, pleasure and thy golden sleep? [loss of interest, insomnia]
Why dost thou bend thine eyes upon the earth, [depression]
And start so often when thou sit'st alone? [hyperactive/startle reaction]
Why hast thou lost the fresh blood in thy cheeks;
And given my treasures and my rights of thee
To thick-eyed musing and curst melancholy?
In thy faint slumbers I by thee have watch'd, [fragmented, vigilant sleep]
And heard thee murmur tales of iron wars; [traumatic dreams, reliving episodes of combat]
Speak terms of manage [horsemanship] to thy bounding steed;
Cry 'Courage! to the field!' And thou hast talk'd
Of sallies and retires, of trenches, tents,
Of palisadoes, frontiers, parapets,
Of basilisks, of cannon, culverin,
Of prisoners' ransom and of soldiers slain,
And all the currents of a heady fight.
Thy spirit within thee hath been so at war,
And thus hath so bestirr'd thee in thy sleep,
That beads of sweat have stood upon thy brow [night sweats]
Like bubbles in a late-disturbed stream;
And in thy face strange motions have appear'd,
Such as we see when men restrain their breath
On some great sudden hest. O, what portents are these?
Some heavy business hath my lord in hand,
And I must know it, else he loves me not.

Observe in particular the fourth line of the soliloquy (Altschuler, 2018): "Tell me, sweet Lord, what is't that takes from thee/Thy stomach, pleasure, and thy golden sleep." With one wonderful and still novel today contraction, "is't," Shakespeare highlights that PTSD (1) is a distinct nosological entity and (2) one that is not understood. In other words, PTSD is not just melancholia, alcohol or substance abuse, madness, or some combination of these. This is even more remarkable as not only the DSM and modern psychiatry, but also the concept of nosology lay centuries in the future!

As PTSD seems is a ubiquitous accompanier of war perhaps, we might be able to find PTSD depicted in art from past times. Given the scale of the carnage of the US Civil War—more than 600,000 killed out of a population of 38 million—and the brutal nature of the battles which were often fought at close quarters as per 18th century military doctrine, but with far more advanced weaponry, that might be a place to look for PTSD in art.

Indeed it is. The famed American painter Winslow Homer was born in Boston in 1836. His mother, to whom Winslow was close throughout his life, was quite a good amateur watercolorist. His father was a businessman who left the hardware store business to seek his fortune in the California gold rush. Homer was sent by Harper's Weekly as an illustrator to the front in 1861—today we would say he was embedded with the Army of the Potomac.

Homer's first painting, Sharpshooter (Fig. 4.1 1863, now at the Portland Museum of Art), was a remake of one of his illustrations. In a letter to a friend years after the war, Homer wrote, "I looked through one of their

Figure 4.1 Sharpshooter. Winslow Homer's sharpshooter (1863, Portland Museum of Art).

rifles once .... The ... impression struck me as being as near murder as anything I could think of in connection with the army and I always had a horror of that branch of the service." I saw the painting when it was on loan to an exhibit on the Civil War at the Metropolitan Museum of Art in New York City.

It is a wonderful and striking picture.

As time passes from the US Civil War, fewer and fewer people know what someone in a blue uniform is or means, so I was able to view the painting in silence and solitude. It is loudly and violently disturbing in its peace and silence: to think of the person, yes another person, on the other end of the bullet about to die and not even have any idea whence or by whom his life just ended.

The Met's permanent collection includes another Homer masterpiece, his 1865 Veteran in a New Field (Fig. 4.2). The vet has turned his back on society—or is it on humanity or life? While the harvest represents the cycle of life, the scythe certainly evokes thoughts of the reaper.

Homer depicted PTSD, and one wonders if in fact he himself suffered from PTSD secondary to what he witnessed while reporting on the Civil War.

While as illustrator and painter Homer was a master of scenes depicting men, probably like most or all great artists, Homer wanted to portray the timeless and eternal. His paintings from later in life, in of scenes of Nature,

Figure 4.2 Veteran in a new field (Winslow Homer, 1865; Metropolitan Museum of Art).

Figure 4.3 Northeaster (Winslow Homer, 1895/1901, Metropolitan Museum of Art).

especially his maritime scenes such as his classic Northeaster (Fig. 4.3), certainly achieve this goal.

Examples of PTSD in sexual trauma survivors have not yet been found in historical or artistic sources. A search for such examples is warranted.

Recently, I found (Altschuler, 2016) a previously undescribed form of PTSD, one that can have important consequences for our modern society, and also likely did in the past. I found the case in the memoir of Robert Gates (Gates, 2014). In late 2011, Gates abruptly resigned as United States Secretary of Defense after serving in the position for four and a half years under two Presidents of different political parties. In the Author's Note to his recently published memoir, Duty, Gates comments, "Toward the end of my time in office, I could barely speak to them [soldiers] or about them without being overcome with emotion. Early in my fifth year, I came to believe my determination to protect them … was clouding my judgment and diminishing my usefulness to the president …." In the book Gates noted that "the hardest part of being secretary for me was visiting the wounded in hospitals … and it got harder each time." In reflecting on the second to last page of the book Gates writes, "… in my mind's eye I could see them [injured soldiers] lying awake, alone, in the hours before dawn, confronting their pain and their broken dreams and shattered lives. I would wake in the night, think back to a wounded soldier or Marine I had seen … and in my imagination, I would put myself in his hospital room and I would hold him to my chest, to comfort him … so my answer to the young soldier's question … about what kept me awake at night: he did."

We see that Gates has ongoing and longstanding recurrent recollections, dreams, and awakenings about the injuries of soldiers under his command. These caused a significant vocational disturbance—he had to retire from his job! If these recollections were about trauma that he had personally experienced, then Gates would meet the criteria for PTSD. As best is known, Gates personally did not experience such trauma firsthand, so his feelings and emotions are solely induced by trauma sustained by his subordinates. Gates' form of PTSD is unlike PTSD by proxy in caregivers or loved ones of the individual who experienced a trauma because, in this case, PTSD is being experienced by the person who ordered the traumatized individuals into the situation that induced the trauma.

PTSD induced by trauma of subordinates has likely been experienced by others, either as described here, or as a mixed form induced also by personal trauma. This form of PTSD is likely an important workplace hazard for civilian and military war commanders. This occupational hazard may select for leaders who are callous or are inured to the effects of it.

References

Abdul-Hamid, W. K., & Hughes, J. H. (2014). Nothing new under the sun: Post-traumatic stress disorder in the ancient world. *Early Science Medicine, 19*(6), 549–557.

Altschuler, E. L. (2016). PTSD induced by the trauma of subordinates: The Robert Gates syndrome. *Occupational Medicine, 66*(3), 182. https://doi.org/10.1093/occmed/kqv201

Altschuler, E. L. (2017). *A great literature guide to the DSM-5.* Oxford.

Altschuler, E. L. (November 2018). *Lessons and guidance for contemporary neuroscience from historical sources: Examples from Shakespeare and Bach. Poster session J 021.02.* San Diego, CA: Annual Meeting of the Society for Neurosciences. https://www.abstractsonline.com/pp8/#!/4649/presentation/4961.

Bentley, S. (2005). *Short history of PTSD: From Thermopylae to Hue soldiers have always had a disturbing reaction to war. VA Veteran* http://archive.vva.org/archive/TheVeteran/2005_03/feature_HistoryPTSD.htm.

Clawson, A. H., Jurbergs, N., Lindwall, J., & Phipps, S. (2013). Concordance of parent proxy report and child self-report of posttraumatic stress in children with cancer and healthy children: Influence of parental posttraumatic stress. *Psycho-Oncology, 22*(11), 2593–2600. https://doi.org/10.1002/pon.3321

Croq, M. A., & Croq, I. (2000). From shell shock and war neurosis to posttraumatic stress disorder: A history of psychotraumatology. *Dialogues in Clinical Neuroscience, 2*(1), 47–55. https://doi.org/10.31887/dcns.2000.2.1/macrocq

Da Costa, J. M. (1871). On irritable heart: A clinical study of a form of functional cardiac disorder and its consequences. *American Journal of the Medical Sciences, 61*(121), 17–52.

Erichsen, J. E. (1867). *On railway and other injuries of the nervous system.* Henry C. Lea.

Ezra, M. B. (2001). Earliest evidence of post-traumatic stress? *British Journal of Psychiatry, 179*(5). https://doi.org/10.1192/bjp.179.5.467, 467–467.

Fluckiger-Hawker, E. (1999). *Urnamma of Ur in sumerian literary tradition.* University Press Fribourg.

Gates, R. (2014). *Duty.* Vintage.

Grinker, R. R., & Spiegel, J. P. (1945). *Men under stress*. Blakiston.
Huber, T. J., & Te Wildt, B. T. (2005). Charles Dickens' a tale of two cities: A case report of posttraumatic stress disorder. *Psychopathology, 38*(6), 334–337. https://doi.org/10.1159/000089454
Jayatunge, R. M. (2010). *Post-traumatic stress disorder (PTSD) – a malady shared by East and west: A Sri Lankan look at combat stress and trauma*. https://community.apan.org/wg/tradoc-g2/fmso/m/fmso-monographs/239222.
Kaufmann, F. (1916). Die planmaßige Heilung komplizierter psychogener Bewe- gungs-storungen bei Soldaten in einer Sitzung. *Feldartz Beilage Munch Med Wochenschr, 63*, 802ff.
Milian, G. (1915). L'hypnose des batailles. *Paris Medicine*, 265–270.
Myers, C. (1915). A contribution to the study of shell shock. *The Lancet, 185*(4772), 316–320. https://doi.org/10.1016/s0140-6736(00)52916-x
Salmon, T. W. (1917). Care and treatment of mental disease and war neuroses (shell shock) in the British army. *Metal Hygiene, 1*, 5–9.
Shay, J. (1994). *Achilles in Vietnam*. Scribner.

SECTION 2

Applications of AAT

CHAPTER 5

The use of service dogs for military veterans with posttraumatic stress disorder

James Whitworth
University of Central Florida, Orlando, FL, United States

Introduction

An estimated 11%–20% of military veterans in the United States who served in recent conflicts are dealing with posttraumatic stress disorder (PTSD) related to their time in the service, and about 30% of Vietnam Veterans experience this disorder during their lifetime (National Center for PTSD, 2018). An additional 2.3%–22.3% of military members and veterans experience subdiagnostic levels of PTSD but don't meet all the criteria for the disorder (Bergman et al., 2017). Service dogs are increasingly being employed to help military veterans recover from traumatic experiences that occurred during their time in the military.

The use of service dogs provides veterans dealing with trauma responses with a viable complementary or alternative treatment to more traditional treatment modalities. Service dogs can support these veterans who engage in current research-based interventions such as prolonged exposure therapy (PET), eye movement desensitization and reprocessing (EMDR), or cognitive processing therapy (CPT). While PET, EMDR, and CPT are fully supported by the Department of Veterans Affairs (VA) and the Department of Defense, about half of all veterans with PTSD either decline to participate in these interventions or drop out of them early on in the treatment process (Deviva et al., 2016; Goetter et al., 2015; Hoge et al., 2014; Monson et al., 2006; Peterson et al., 2011, p. 42–58). By contrast, veterans with PTSD are more willing to start and finish service dog-related programs to help them address their symptoms (Whitworth et al., 2019).

Animal Assisted Therapy Use Application by Condition
ISBN 978-0-323-98815-5
https://doi.org/10.1016/B978-0-323-98815-5.00005-7

Description of service dogs for veterans with PTSD

A service dog is "a dog that is individually trained to do work or perform tasks for a person with a disability" (American Kennel Club, 2019). There are an estimated 17,000 service dogs trained by Assistance Dogs International (ADI)-accredited organizations in the United States and Canada (Assistance Dogs International, 2020). This number, however, excludes the many thousands of dogs that are trained by their owners to work as their service dogs. Service dogs are different from emotional support dogs and canines that serve as companions or pets. While these other types of dogs can be incredibly supportive, they do not provide the many highly distinctive tasks that service dogs perform for these veterans. Emotional support dogs offer comfort to help their owners deal with a mental health condition. Service dogs provide much more than emotional support or companionship. They are trained to accomplish particular responsibilities for someone who cannot complete those functions because of their physical and/or psychological disability. A service dog, for example, can help someone with a seizure disorder by alerting them to an oncoming seizure or keeping them safe during these episodes. As an example, for an individual dealing with visual impairment or blindness, a service dog can aid in their accomplishing the task of traversing carefully down a street. Service dogs help individuals with tasks that they often can't complete without some assistance, as a result of limitations, and such tasks are usually distinct from natural dog behavior. They are generally focused on maintaining safety and improving their owner's global functioning as they help them with a broad range of diverse tasks based on their specific needs.

Service dogs perform specific functions for veterans who struggle accomplishing tasks because of them having PTSD. They can aid them by alerting them to potential triggers, helping them manage avoidance behaviors, and decreasing their hyperarousal, hypervigilance, and reexperiencing of their trauma along with assisting them to complete specific behaviors when symptoms are present (Crowe et al., 2018; Rodriguez et al., 2018; Whitworth et al., 2019). When needed, these service dogs can also assist veterans with tasks such as reminding them to take their medications, retrieving dropped items, or keeping their balance (Whitworth et al., 2019). They can also provide veterans with a sense of purpose and help them reconnect with others in their lives especially spouses, partners, children, parents, friends, and coworkers (Crowe et al., 2018; Rodriguez et al., 2018). Service dogs can help veterans manage their PTSD symptoms by interrupting their anger episodes or hypervigilance, waking them from a nightmare, helping them be present in the moment, or sensing their hyperarousal and then getting their attention (Whitworth et al., 2019).

Service dog training programs for veterans with PTSD

A broad range of nonprofit and for-profit organizations across the United States provide and train service dogs for veterans dealing with PTSD. These agencies are either community-based where they primarily focus on the veterans in their general vicinity, while others employ a national program model where they provide service dogs to veterans from anywhere in the country. Some agencies only train purpose-bred dogs for this role, while others will train shelter/rescue and companion dogs to become service dogs for veterans. Training a service dog for this role commonly takes 6 to 12 months or more to complete depending on the model used by the agency with the cost ranging from $15,000 to $50,000 depending on the type of training (National Service Animal Registry, 2019). This can include training for the veteran who receives the dog and periodic follow-up training for the dog/vet pair to ensure the service dog routines and standards are being maintained. Many organizations provide service dogs to veterans at no cost, while others charge a fee for a trained dog, and some may offer financial aid for those who cannot afford a service dog. Drop-out rates for dogs that are candidates to be service animals can run as high as 50% to 70% due to the strict requirements needed for dogs to function in this role (Atlas Assistance Dogs, 2021). Training is primarily conducted by professional service dog trainers.

There are typically two models for this training. One model has the veteran and canine pair participate jointly in extended weekly classes, along with extensive guidance from a professional trainer, teaching the canine to become their service dog. The second model is where the agency pretrains the canine to become a service dog after which they bring in the veteran for 1–3 weeks of training on how to own and maintain their service dog.

The following description provides details about the training content and practices for a large service dog agency in the southwest United States that provides and trains canines for veterans with PTSD.

The service dog training program consists of weekly 60-minute sessions where veterans are taught to train their dogs in small group classes consisting of no more than 10 participants. These training groups are predominately conducted at the program's facilities, where there is appropriate space to allow for adequate training exercises and practice. All sessions are taught by a Certified Master Dog Trainer who provides individual and group guidance to participants as they train their own canine. Veterans are provided time to socialize with each other before, during, and after each session, and they frequently discuss their common challenges and similar pursuits, primarily related to their dogs, according to program staff. The dog trainer and staff all seek to provide a safe and accepting environment for the veterans. The program curriculum primarily centers on training each veteran and dog pair together to do essential obedience tasks such as how to heel, sit, stay, and lay down. Each participant is trained on how to recognize their own dog's response to their personal needs. Through this process, the training seeks to refocus the veteran's thoughts on the task at hand (i.e., training their dog to do collaborative tasks) instead of on thoughts centered on themselves and any distractions that may lead to increased anxiety. The program aims to help each participant improve their sense of control and mastery over their reactions, emotions, and behaviors, as they increase their skill with their dog. Since the dogs live with the veterans, they are expected to reinforce each week's lesson with their dog on their own between weekly sessions. The concluding session requires participants to take their dogs to a large department or hardware store where they demonstrate their developed skills with the dog. Veteran and dog pairs who are unable to successfully demonstrate each skill will retake the last seven classes in the training. All participants are invited to return to the program at any point to enhance their dog training skills, to work toward dog training certification, to volunteer to assist staff, or to socialize with and support the other veterans who are there. (Scotland-Coogan et al., 2020, p. 9—10).

There are very limited accepted standards of uniform practice that are used by the service dog agencies to provide and train canines for veterans with PTSD. Even though most service dog agencies use certified professional trainers to conduct these programs, many rely principally on the trainer's practice experience and anecdotal perceptions to guide this work. There is also limited research-based guidance on the best sources for canines to be trained as service dogs and how to match or pair service dogs with veterans. Veterans or others wanting to have a service dog are permitted by the Americans with Disabilities Act (ADA) of 1990 to train a service dog themselves are not required to use a professional service dog trainer or training program to have their canine professionally trained (ADA.gov, 2021).

Current guidelines for selecting and pairing service dogs for veterans

The Service Dogs for Veterans National Training Standards developed by the Association of Service Dog Providers for Military Veterans specify that a service dog for a veteran "must be selected with the expectation that it will meet all of the requirements for service animals as defined by Americans with Disabilities Act (ADA) of 1990, and, in addition, be able to cater to the unique needs of Veteran with PTS, TBI, and/or MST." (ADA.gov, 2021, p. 3). They further stipulate that the canine should be healthy, trainable to be a service dog, have limited predilection to bark and/or whine excessively, spayed/neutered by the completion of their training that they should be at least 17 weeks of age before public access training, should not enter training after its third birthday (with exceptions up to four years and for family pets on a case-by-case basis), must be house-broken, and that have all required vaccinations. These standards also note that all potential service dogs must undergo a temperament test to determine whether or not the dog "(1) shows aggression toward people, children, other animals, or other dogs; (2) has low to medium reactivity and/or arousal to sounds, animals, and other stimuli; (3) displays sufficient levels of neoteny; and (4) otherwise displays the friendly and servicing nature of a high-quality Service Dog." (ADA.gov, 2021, p. 3). Concerning the size of the service dog for a veteran with PTSD, these standards also state that they should ideally be 24 inches at the shoulder and at least 60 pounds and no taller than 34 inches at the shoulder and not heavier than 99 pounds with notable exceptions as long as the dog is not too large to practically serve in this role. These standards do not restrict any breed from becoming a service dog, but they do advise against the selection of canines that would limit the veteran's ability to make social connections.

The Association of Service Dog Providers for Military Veterans requires associated agencies to pair service dogs with a veteran based upon the "best fit" for each dog/veteran team with consideration of the veteran's mental and physical abilities and how they match with the dog's temperament. Other considerations for the best fit for pairing include the veteran's family, work/school life, lifestyle, and other responsibilities they may have. Many agencies will conduct home visits to assess the appropriateness and compatibility of the setting for having a service dog living there to include the impacts on other animals and humans living in the home. Most organizations also require that the veteran have a good conduct discharge from the military and that they have not reported a history of violence to humans or animals.

Standards developed by ADI note that dogs who serve in assistance roles should be "temperamentally screened for emotional soundness and working ability, physically screened for the highest degree of good health and physical soundness, and technically and analytically trained for maximum control and for the specialized tasks he/she is asked to perform." (Assistance Dogs International, 2020). ADI requires all of its member organizations to demonstrate a clear and comprehensive application, screening, and matching processes for clients. They also require that dogs "must be matched to best suit the client's needs, abilities, and lifestyle and that they are placed with a client able to interact with him/her." (Assistance Dogs International, 2020). ADI member organizations also agree to not train, place, or certify dogs showing any aggressive behavior with nonaggressive barking being acceptable in some appropriate situations.

Many service dog agencies also use the 10-step Canine Good Citizen (CGC) test from the American Kennel Club (AKC) to assess good manners in dogs and owners alike. This test is used by many of these agencies as a standard that their participating canines must meet before being considered a service dog. This test gauges the dog's ability, as guided by their human partner, to accept a friendly stranger, sit politely, walk on a loose leash, walk through a crowd, sit and lay down on command, stay in place, come when called, and react (or not react) to a distraction or to other dogs. The AKC further notes that a service dog candidate should "be calm, especially in unfamiliar settings, be alert (but not reactive), have a willingness to please, be able to learn and retain information, be capable of being socialized to many different situations and environments, and be reliable in performing repetitive tasks" (American Kennel Club, 2020).

Reported benefits of service dogs for veterans with PTSD

There is limited but mounting research support for the use of specially trained service dogs as a complementary approach to help veterans experiencing PTSD. In a growing number of studies addressing the efficacy of these programs, participating veterans are reporting benefits including decreased PTSD symptoms, anger, depression, suicidal ideations, social isolation, and medication usage. They are also describing improvements in their sleep, relational/social/work functioning, and physical/medical condition. Each of these reported benefits is described in this section along with a discussion of the published research supporting the improvements.

Mental health benefits

Decreased PTSD symptoms: Most efficacy studies into these programs to date have justifiably focused chiefly on the impact they may have in decreasing the participating veteran's PTSD symptoms. Veterans in nine published studies consistently report statistically significant or close to significant decreases in their PTSD symptoms including avoidance of stimuli, intrusive thoughts, negative alterations in cognition and mood, and arousal/reactivity associated with their trauma experiences (Bergen-Cico et al., 2018; Farmer, 2021; Kloep et al., 2017; O'Haire & Rodriguez, 2018; Rodriguez et al., 2020; Scotland-Coogan et al., 2020; Whitworth et al., 2019; Yarborough et al., 2018; Yount et al., 2012). They have noted such improvements in these studies which primarily used the PTSD Checklist for DSM-5 (PCL-5), the PTSD Checklist-Military Version (PCL-M), or the Trauma Symptom Inventory-2 (TSI-2) to assess the veteran's PTSD symptoms (Blevins et al., 2015; Briere, 2011; Weathers et al., 1993). In all of these studies cited here, the decreased symptoms were significant, but they were not associated with a loss of the participant's PTSD diagnosis.

One of the most thorough studies to date in this area used a cross-sectional study design that included 217 veterans with PTSD, $n = 134$ who had been matched with a service dog and $n = 83$ on a waitlist to be matched with a service dog (Rodriguez et al., 2020). The study sought to not only determine the efficacy of the examined program but also identify the trained service dog tasks that were most helpful in reducing the participating veterans' PTSD symptoms. The dog tasks monitored in the study included interrupt/alert to anxiety, calm/comfort anxiety, block (create space), block (guard/protect), cover (watch back), social greeting, and wake up from a nightmare. Researchers in this study concluded that all of these tasks provided by the service dogs, including some of their dog's untrained behaviors in relation to them, helped reduce the participating veteran's PTSD symptoms. Findings showed that the dog's task of calming and interrupting their veteran's anxiety was identified as the most frequently used in a typical day for reducing PTSD symptoms, especially hypervigilance and intrusion.

A qualitative investigation into the impact of these programs conducted semistructured interviews with 21 veterans dealing with PTSD who had received a service dog (Krause-Parello & Morales, 2018). Participants consistently noted that their relationship with their service dog provided them with a means of social support that helped them cope with their

PTSD symptoms that in turn improved their independence. The following quotes from veterans in this study exemplify the role that service dogs can play in helping these veterans with their PTSD symptoms:

"If I have a really bad episode, he'll [service dog] get really close to me and he'll just kind of let me pet him or he'll even jump in my lap and just kind of lick me and tell me it'll be okay, you know? When we're out in public and I have one he'll get me out of the situation, he'll pull to where I can go get calmed down and be okay."

"She [service dog] will wake me up from nightmares, so that way it doesn't get to the point where I start thrashing in my sleep and I don't get violent in my sleep." (Krause-Parello & Morales, 2018, p. 69).

Decreased depression: Veterans participating in service dog programs have reported experiencing less depression and improved mood after receiving and training their canine (Kloep et al., 2017; O'Haire & Rodriguez, 2018; Scotland-Coogan, 2019). The following excerpt from a qualitative investigation that conducted intensive interviews with 15 participants in these programs demonstrates the role that having a service dog may play in addressing the veteran's mood:

The service dogs demonstrate the ability to sense the emotions of the veterans. Throughout the interviews, the service dogs would sleep until the veteran started discussing something which was emotionally taxing. Participant seven shared, "He knows, he knows when I'm upset, you know. And he knows when he needs to sit with me cause right now, he's sitting underneath there and he's looking sad like what's going on here?" When this happened, the dogs would first look up to assess what was happening; if they sensed emotional upset, they would get up and relentlessly seek the veterans' full attention (Scotland-Coogan, 2019, p. 1903).

Decreased anger: Military veterans with PTSD deal with high levels of anger which is often associated with more severe PTSD overall (Kulkarni et al., 2012). Some of the investigations into service dog programs for these veterans have found that participants and their family members report that the veterans demonstrate notably decreased levels of anger after receiving and owning their canine (Kloep et al., 2017; Rodriguez et al., 2018; Whitworth et al., 2020). One partner of a veteran who described her husband as dealing with a great amount of rage related to his deployment experiences described the service dog's impact on their veteran's anger in the following way "So, he now has [service dog] and [service dog] will sort of be like, 'I sense something is about to happen. Let's go somewhere else,' or 'It's time to sit down' or 'Let's find a quiet spot'." (Whitworth et al., 2020, p. 12).

Improved sleep: Sleep impairments are among some of the most prevalent and destructive symptoms that veterans with PTSD deal with. Nearly half of all veterans with deployment-related PTSD also have difficulties obtaining restorative and beneficial sleep (Khazaie et al., 2016; Liempt et al., 2013). Only a small number of studies to date have sought to determine what impact involvement in a service dog program may have on the participating veteran's sleep quality and quantity. One study that compared the experiences of 45 veterans with PTSD who had a service dog to 28 similar veterans on a waitlist to receive one found that those with a dog reported significantly better sleep compared to those waiting to receive a dog (Rodriguez et al., 2018). In a qualitative investigation that included 41 veterans with service dogs and their caregivers, participants consistently reported improvements in sleep quality and duration (Yarborough et al., 2018). In a different study, one veteran dealing with PTSD and associated sleep difficulties described his dog's impact on his sleep in the following way "Before I had a service dog, I needed to take antianxiety medications, I was on antidepressants, I was taking two or three different meds to go to sleep. And now the only thing I'm on is (medication name), in order to help me get to sleep at night." (Krause-Parello & Morales, 2018, p. 69).

Decreased suicidal ideations: It is well known that suicide and suicidal ideations among veterans with PTSD are a prevalent and persistent problem. Two investigations found that veterans reported decreased suicidal ideations after receiving a service dog and completing the associated program (Scotland-Coogan et al., 2020; Yarborough et al., 2018). Very few studies, however, have attempted to identify if having a service dog can decrease suicidal thoughts or behaviors. This is likely due in part to the difficulty in accurately and validly measuring suicidal ideations or actions in this population. Veterans and their partners, on the other hand, regularly describe their service dog as "saving their lives" because they had often reported that they had contemplated or attempted suicide before obtaining their dog. Once such veteran described his thoughts and actions regarding suicide before getting the service dog in the following words, "I probably tried to commit suicide a good 200 times. I'd practice with my nine mil and stick it in my mouth. Without a bullet in it, I would practice pulling the trigger." (Krause-Parello & Morales, 2018, p. 68). The partner of one veteran described the role the service dog had on her spouse's life in the following way "He [the veteran] feels like someone has his back. … 'Got his six'. … the biggest thing is she saves his life. She is his companion." (Whitworth et al., 2020, p. 10).

Improvements in relational/social/occupational functioning

Many veterans who have received a service dog report that their canine has greatly impacted their willingness and ability to be around other people outside of their immediate family members (Bergen-Cico et al., 2018; Krause-Parello & Morales, 2018; O'Haire & Rodriguez, 2018; Rodriguez et al., 2020; Scotland-Coogan, 2019; Scotland-Coogan et al., 2020; Whitworth et al., 2019). PTSD among military veterans is often associated with increased social isolation and difficulty relating with others including spouses/partners. Some are particularly hesitant to relate to individuals who are not connected to the service or to be around those who were not part of the military unit they previously served with. These difficulties can also often impair the veteran's ability to obtain and maintain employment. Improvements in this area can therefore be especially impactful in the veteran's life since they allow them to reconnect with family members and friends and in some cases return to function on their jobs. One veteran described the impact their dog has on their ability to be social with others by stating "Oh, lord, um, [laughter] I'm not in my house anymore ... I get out and socialize ... living a life I didn't think I'd have again. Um it-it's- I can socialize. I can go to church again. You know it- I was so mad at God but, you know, I know I'll get over that and I go to church, I socialize, I do my own shopping. It's like I have my own life back." (Scotland-Coogan, 2019, p. 1904).

Physical/medical impacts

Little is known about the impacts that having a service dog might have on a veteran's physical and medical status. Many veterans and their caregivers who participated in one service dog program repeatedly stated during in-depth interviews that the veteran was using fewer prescription medications due to having a service dog (Yarborough et al., 2018). A separate study used a cross-sectional investigation design to quantify the effects of having a service dog on medication use by US Post 9/11 veterans with PTSD (Rodriguez et al., 2021). When comparing 52 veterans with a service dog to 44 on a waitlist to receive a one, they found no significant effects of having a service dog on overall self-reported use of medications. Veterans with a service dog, however, did report that their medical provider had reduced their dosage or removed medications after they received their dog. Another study specifically examined the effect of having a service dog on the salivary cortisol awakening response (CAR) and arousal-related functioning in post-911 military veterans with PTSD (Rodriguez et al., 2018). They found that veterans in the study who had a service dog had higher levels of CAR and arousal-related functioning which reflected better health and well-being among these veterans compared to those on a waitlist to receive a dog.

Impacts of service dogs on family members of veterans with PTSD

Family members can be greatly impacted by being connected to a veteran with PTSD. Spouses or partners of previously deployed veterans with PTSD consistently have higher rates of psychological and marital distress compared to those who are in a relationship with a veteran who does not have PTSD (Monson et al., 2009; Renshaw et al., 2008). They can experience secondary traumatic stress (STS) in response to being with and caring for veterans dealing with PTSD (Ahmadi et al., 2011; Bride & Figley, 2009). STS responses frequently include sleep disturbances, avoidance symptoms, emotional regulation challenges, and negative beliefs about self, others, and the world (Bride & Figley, 2009). Given the reported benefits for veterans when they have a service dog, it is important to know how the dog along with associated changes in these veterans may impact their family members. Initial studies to date have shown that having a service dog is largely beneficial for partners, spouses, and family members (Bibbo et al., 2019; McCall et al., 2020; Whitworth et al., 2020). Based on a comparison of responses from caregivers/spouses of veterans with PTSD who had a service dog, to responses from those without a dog, one study concluded that having a service dog was associated with them having lower levels of health-related worry and healthier overall psychosocial and emotional functioning (Bibbo et al., 2019). Another investigation similarly found significant improvements in partner/family and relationship functioning when a veteran had a service dog in contrast to those without one (McCall et al., 2020). When asked to describe how the presence of a service dog in their family may have changed things for their relationships, one spouse stated "With [dog], [veteran] has been more able to come out and socialize even with just the family, instead of isolating.... He used to isolate in the bedroom and not be able to handle it, but with [dog] with him, he is able to come out and spend more time with the family and interact and participate." (Whitworth et al., 2020, p. 10). Even though having a service dog will often necessitate adjustments for the entire partner/family system, many veterans have made comments like the following about having their dog as part of these relationships "We're closer [with the dog]. We talk more. Especially if we have, we don't fight but if we have a petty argument, like we never have screaming and yelling matches. We both just agree they have no purpose. We just have, we argue. We don't yell at each other. And if we get too frustrated, we just walk away and when we calm down, we come back ... But we don't argue as much. We sit there and talk things out." (Scotland-Coogan, 2019, p. 1908).

US Department of Veterans Affairs study of service dogs for veterans with PTSD

In perhaps the most rigorous study to date into these programs, the US Department of VA conducted a multiyear multisite investigation that compared the benefits of pairing a service dog versus pairing an emotional support dog with veterans dealing with PTSD (U.S. Department of Veterans Affairs, 2020). The study included 97 veterans with PTSD that were matched with a service dog and 84 veterans that were matched with an emotional support dog. Dogs involved in the study were provided by three different dog agencies from across the country. The study defined service dogs as "a dog that is trained to perform one or more tasks to mitigate a disability" and emotional support dogs as "a dog trained in obedience commands but not trained to perform a task that mitigates a disability" (U.S. Department of Veterans Affairs, 2020, p. 8–9). Although results of this investigation showed that both types of dogs can equally support these veterans in addressing their overall sense of disability and improving their quality of life, service dogs were notably better compared to emotional support dogs in helping the veterans in the study to manage their PTSD symptoms and in decreasing their suicidal behaviors and ideations. The findings of the study highlight some promising and distinct benefits that service dogs appear to provide for these veterans.

Challenges in studying the use of service dogs for veterans with PTSD

Despite growing initial evidence supporting the use of service dogs for veterans with PTSD, major questions regarding their exact potential efficacy, the impact of confounding and interacting independent variables, and how best to utilize canines with this population remain unanswered (van Houtert et al., 2018). Many of the studies into this area to date lack sufficient research designs to control for the many intervening variables involved in conducting these programs. Even the more rigorous trial conducted by the VA that was described earlier in this chapter failed to fully address the issue that decreases in PTSD symptoms reported by veterans participating in their study could be the result of ongoing PTSD treatment unrelated to the study intervention.

These challenges are further compounded by the lack of standardization of practices among the organizations providing and training service dogs for veterans and their heavy reliance on anecdotal information which is often based solely on their own practice experience and subjective perceptions. There are relatively few accepted standards of uniform practice that are used by the service dog agencies that provide and train canines for veterans. Among a long list of unanswered questions, little remains known regarding the optimal training techniques, the best sources to obtain dogs as candidates to be trained as service dogs (i.e., purpose-bred, shelter/rescue, or existing companion), or the best strategies for selecting, matching/pairing a service dog with a veteran. It is important to note that sometimes the pairing of a human with a service dog is unsuccessful which can precipitate negative consequences for everyone involved (Lloyd et al., 2016). The individual may be hesitant to try another match, and the dog may have a negative bonding experience making it difficult for them to be paired with another human (Fallani et al., 2007; Lloyd et al., 2016). Other distinct challenges that may limit the use of this intervention for some veterans with PTSD include the extensive amount of ongoing work required to train and maintain a service dog and the reality that having one can often increase unwanted public attention for the veteran (Yarborough et al., 2018).

This area of study requires rigorous randomized controlled trials that compare veterans who receive service dogs with demographically similar veterans who do not have one or who are participating in an alternative/placebo intervention (Yarborough et al., 2018). These trials "should account for self-selection bias and some veterans' overly optimistic expectations for service dogs, both of which have not previously been controlled for in nonrandomized studies to date and may present a challenge to trial validity, and a long-term follow-up evaluation period should investigate the specific mechanisms of action at work (i.e., how do service dogs improve PTSD symptoms and quality of life)" (Yarborough et al., 2018, p. 118).

In their report entitled "Service Dogs and Wounded Warriors: The State of the Science," Benbassat et al. (2021) make several key recommendations required to improve our understanding of the efficacy and best practices in this field (Benbassat et al., 2021). Their recommendations serve as a highly needed and appropriate agenda for future research into the use of service dogs for veterans with PTSD:

1. Standardize training requirements and competencies.
2. Isolate the effect of the service dog task from the effect of the dog itself (in order to help limit threats to internal validity, external validity, and reliability).
3. Treat waitlist and treatment groups alike, save service dogs (Service dog treatment should be compared to other PTSD interventions and companion dogs).
4. Match control participants with companion or therapy dogs.
5. Assess service dog benefits throughout the lifespan of the dog.
6. Adopt a transformational, as opposed to transactional, approach.

Theoretical underpinnings of service dog programs

A clearly expressed and endorsed theoretical basis for the use of service dogs to help veterans with PTSD has not been fully formulated. These programs emerged as an alternative or complementary approach to standard office-based clinical and medical interventions prior to the development of an underlying theoretical foundation (Esposito et al., 2011). Unlike most clinical interventions for PTSD which rely on cognitive and/or behavior theories, service dog programs do not include any formal effort to encourage participants to emotionally process their traumatic experience(s) or confront pathological fear structures (Briere & Scott, 2015). Animal-assisted therapy predominately uses attachment theory as created by Bowlby (1969) and then supplementarily developed by Ainsworth and Bowlby (1991) as its fundamental theoretical framework (Vitztum & Urbanik, 2016). From an attachment theory perspective, close relationships with consistent and stable humans provide the means primary for humans to develop in healthy ways and for them to deal with and be resilient to trauma experiences. Such attachment relationships, or more specifically the ability to derive comfort from another human being, are the foremost predictors of a person's capacity to recover from traumatic experiences (Johnson, 2002).

Service dogs provide veterans who have experienced PTSD with a new attachment relationship. This new relationship notionally appears to provide an attachment relationship to help these veterans start to experience some comfort from which they might start to process their trauma experience(s) in an experiential but nonconfrontational manner. The veteran's attachment relationship with their service dog, where they can again begin to feel warmth and comfort, may function as a way of helping them reconnect with other people in their lives (O'Haire & Rodriguez, 2018). The presence of their service dog especially when dealing with intrusive thoughts or triggers and reminders of their military trauma can also provide them with a reassuring reminder that they are not currently in a threat environment (Zilcha-Mano et al., 2011).

Using attachment theory as a primary guide, some theorize that having a service dog provides an essential "relational bridge" to veterans with PTSD to bounce back from their traumatic experiences (Whitworth et al., 2020). This relational bridge "is comprised of three core components; (1) forming/maintaining an attachment bond with their service dog; (2) experiencing camaraderie with fellow veterans and others in these programs; and (3) having a close and committed connection with their partner" (Whitworth et al., 2020, p. 611). If the veteran does not have an intimate partner, the third component can alternatively be having close and committed connections with others in their lives. Such a relational bridge may help these veterans foster more resilience to help them manage ongoing challenges resulting from military-related trauma experience(s). Through experiencing a relational connection with their service dog, they may then experience a greater sense that they are no longer in an environment as threatening as they were in during their deployment(s). This bridge may be able to help them focus on their "new mission" and on their own personal, relational, and occupational goals.

Conclusion

The use of service dogs has emerged as a popular complementary intervention to help veterans with PTSD to manage their symptoms and reconnect with others in their lives. Veterans actively seek out and complete service dog programs at much higher rates compared to standardized office-based clinical interventions for military-related PTSD. Having a service dog is associated with lower levels of PTSD symptoms, anger, depression, suicidal ideations, social isolation, and medication usage along with improved sleep, relational/social/work functioning, and physical/medical condition for veterans with PTSD. Having a service dog may provide veterans dealing with PTSD with an attachment-based relational bridge to help them feel more in control of their circumstances and start the process of recovering from their military-related trauma experiences. Research into these programs is just beginning with little still known about the efficacy of this intervention or the identification of standardized best practices. There are many unanswered questions remaining regarding their value and the most effective ways to conduct them. Several key practical research challenges could hinder conducting the use of needed trials in this area. However, rigorous randomized controlled trials still need to be conducted comparing veterans who receive service dogs with those who have not obtained one before the actual efficacy of these programs is better understood.

References

ADA.gov. (2021). *Frequently asked questions about service animals and the ADA*. www.ada.gov/regs2010/service_animal_qa.html.

Ahmadi, K., Azampoor-Afshar, S., Karami, G., & Mokhtari, A. (2011). PTSD research: The association of veterans' PTSD with secondary trauma stress among veterans' spouses. *Journal of Aggression, Maltreatment and Trauma, 20*(6), 636–644. https://doi.org/10.1080/10926771.2011.595761

Ainsworth, M. S., & Bowlby, J. (1991). An ethological approach to personality development. *American Psychologist, 46*(4), 333–341. https://doi.org/10.1037//0003-066x.46.4.333

American Kennel Club. (2019). *Service dog training 101*. www.akc.org/expert-advice/training/service-dog-training-101/.

American Kennel Club. (2020). *Step-by-step CGC training in your home & backyard*. www.akc.org/expert-advice/training/step-step-cgc-training/.

Assistance Dogs International. (2020). *Summary of standards*. https://assistancedogsinternational.org/clientuploads/Summary%20of%20Standards/Summary_ADI_Standards_2020_revision.pdf.

Atlas Assistance Dogs. (2021). *Must read service dog statistics 2021*. www.atlasdog.org/2021/09/23/must-read-service-dog-statistics-2021/?utm_source=rss&utm_medium=rss&utm_campaign=must-read-service-dog-statistics-2021.

Benbassat, D., Buccellato, K. H., Nordstrom, M. J., Jaskot, E. M., Hager, N., & Pasquina, P. F. (2021). Service dogs and wounded Warriors: The state of the science. *The United States Army Medical Department Journal* [Manuscript submitted for publication].

Bergen-Cico, D., Smith, Y., Wolford, K., Gooley, C., Hannon, K., Woodruff, R., Spicer, M., & Gump, B. (2018). Dog ownership and training reduces post-traumatic stress symptoms and increases self-compassion among veterans: Results of a longitudinal control study. *Journal of Alternative and Complementary Medicine, 24*(12), 1166–1175. https://doi.org/10.1089/acm.2018.0179

Bergman, H. E., Przeworski, A., & Feeny, N. C. (2017). Rates of subthreshold PTSD among U.S. Military Veterans and Service members: A literature review. *Military Psychology, 29*(2), 117–127. https://doi.org/10.1037/mil0000154

Bibbo, J., Rodriguez, K. E., & O'Haire, M. E. (2019). Impact of service dogs on family members' psychosocial functioning. *American Journal of Occupational Therapy, 73*(3). https://doi.org/10.5014/ajot.2019.031690

Blevins, C. A., Weathers, F. W., Davis, M. T., Witte, T. K., & Domino, J. L. (2015). The posttraumatic stress disorder checklist for DSM-5 (PCL-5): Development and initial psychometric evaluation. *Journal of Traumatic Stress, 28*(6), 489–498. https://doi.org/10.1002/jts.22059

Bowlby, J. (1969). *Attachment and loss* (Vol. 1). Basic Books.

Bride, B. E., & Figley, C. R. (2009). Secondary trauma and military veteran caregivers. *Smith College Studies in Social Work, 79*(3–4), 314–329. https://doi.org/10.1080/00377310903130357

Briere, J. (2011). Trauma symptom inventory—2 (TSI—2). *Psychological Assessment Resources.*

Briere, J., & Scott, C. (2015). *Principles of trauma therapy: A guide to symptoms, evaluation, and treatment.* Sage Publications.

Crowe, T. K., Nguyen, M. T., Tryon, B. G., Barger, S., & Sanchez, V. (2018). How service dogs enhance veterans' occupational performance in the home: A qualitative perspective. *The Open Journal of Occupational Therapy, 6*(3). https://doi.org/10.15453/2168-6408.1468

Deviva, J. C., Sheerin, C. M., Southwick, S. M., Roy, A. M., Pietrzak, R. H., & Harpaz-Rotem, I. (2016). Correlates of VA mental health treatment utilization among OEF/OIF/OND veterans: Resilience, stigma, social support, personality, and beliefs about treatment. *Psychological Trauma: Theory, Research, Practice, and Policy, 8*(3), 310–318. https://doi.org/10.1037/tra0000075

Esposito, L., McCardle, P., McCune, S., & Griffin, J. A. (2011). Introduction. In P. McCardle, S. McCune, J. A. Griffin, L. Esposito, & L. Freund (Eds.), *Animals in our lives: Human-animal interaction in family, community, and therapeutic settings.* Paul H. Brookes Publishing Company.

Fallani, G., Prato Previde, E., & Valsecchi, P. (2007). Behavioral and physiological responses of guide dogs to a situation of emotional distress. *Physiology & Behavior, 90*(4), 648–655. https://doi.org/10.1016/j.physbeh.2006.12.001

Farmer, S. M. (2021). Integrating animal-assisted interventions in United States veterans with self-reported posttraumatic stress disorder. *Journal for Nurse Practitioners, 17*(5), 619–622. https://doi.org/10.1016/j.nurpra.2021.02.001

Goetter, E. M., Bui, E., Ojserkis, R. A., Zakarian, R. J., Brendel, R. W., & Simon, N. M. (2015). A systematic review of dropout from psychotherapy for posttraumatic stress disorder among Iraq and Afghanistan combat veterans. *Journal of Traumatic Stress, 28*(5), 401–409. https://doi.org/10.1002/jts.22038

Hoge, C. W., Grossman, S. H., Auchterlonie, J. L., Riviere, L. A., Milliken, C. S., & Wilk, J. E. (2014). PTSD treatment for soldiers after combat deployment: Low utilization of mental health care and reasons for dropout. *Psychiatric Services, 65*(8), 997–1004. https://doi.org/10.1176/appi.ps.201300307

van Houtert, E. A. E., Endenburg, N., Wijnker, J. J., Rodenburg, B., & Vermetten, E. (2018). The study of service dogs for veterans with post-traumatic stress disorder: A scoping literature review. *European Journal of Psychotraumatology, 9*(Suppl. 3), 1503523. https://doi.org/10.1080/20008198.2018.1503523

Johnson, S. (2002). *Emotionally focused couples therapy with trauma survivors: Strengthening attachment bonds.* Guilford Press.

Khazaie, H., Ghadami, M. R., & Masoudi, M. (2016). Sleep disturbances in veterans with chronic war-induced PTSD. *Journal of Injury & Violence Research, 8*(2), 99–107. https://doi.org/10.5249/jivr.v8i2.808

Kloep, M. L., Hunter, R. H., & Kertz, S. J. (2017). Examining the effects of a novel training program and use of psychiatric service dogs for military-related PTSD and associated symptoms. *American Journal of Orthopsychiatry, 87*(4), 425–433. https://doi.org/10.1037/ort0000254

Krause-Parello, C. A., & Morales, K. A. (2018). Military veterans and service dogs: A qualitative inquiry using interpretive phenomenological analysis. *Anthrozoos, 31*(1), 61–75. https://doi.org/10.1080/08927936.2018.1406201

Kulkarni, M., Porter, Katherine E., & Rauch, S. A. M. (2012). Anger, dissociation, and PTSD among male veterans entering into PTSD treatment. *Journal of Anxiety Disorders, 26*(2), 271–278. https://doi.org/10.1016/j.janxdis.2011.12.005

Liempt, S., Zuiden, M., Westenberg, H., Super, A., & Vermetten, E. (2013). Impact of impaired sleep on the development of PTSD symptoms in combat veterans: A prospective longitudinal cohort study. *Depression and Anxiety, 10.*

Lloyd, J., Budge, C., La Grow, S., & Stafford, K. (2016). An investigation of the complexities of successful and unsuccessful guide dog matching and partnerships. *Frontiers in Veterinary Science, 3.* https://doi.org/10.3389/fvets.2016.00114

McCall, C. E., Rodriguez, K. E., MacDermid Wadsworth, S. M., Meis, L. A., & O'Haire, M. E. (2020). "A part of our family"? Effects of psychiatric service dogs on quality of life and relationship functioning in military-connected couples. *Military Behavioral Health, 8*(4), 410–423. https://doi.org/10.1080/21635781.2020.1825243

Monson, C. M., Schnurr, P. P., Resick, P. A., Friedman, M. J., Young-Xu, Y., & Stevens, S. P. (2006). Cognitive processing therapy for veterans with military-related posttraumatic stress disorder. *Journal of Consulting and Clinical Psychology, 74*(5), 898–907. https://doi.org/10.1037/0022-006X.74.5.898

Monson, C. M., Taft, C. T., & Fredman, S. J. (2009). Military-related PTSD and intimate relationships: From description to theory-driven research and intervention development. *Clinical Psychology Review, 29*(8), 707–714. https://doi.org/10.1016/j.cpr.2009.09.002

National Service Animal Registry. (2019). *How much does a service dog cost.* www.nsarco.com/blog/service-dog-buyers-guide.html.

National Center for PTSD. (2018). *How common is PTSD in veterans?.* www.ptsd.va.gov/understand/common/common_veterans.asp.

O'Haire, M. E., & Rodriguez, K. E. (2018). Preliminary efficacy of service dogs as a complementary treatment for posttraumatic stress disorder in military members and veterans. *Journal of Consulting and Clinical Psychology, 86*(2), 179–188. https://doi.org/10.1037/ccp0000267

Peterson, A. L., Foa, E. B., & Riggs, D. S. (2011). *Treating PTSD in military personnel: A clinical handbook.* The Guilford Press.

Renshaw, K. D., Rodrigues, C. S., & Jones, D. H. (2008). Psychological symptoms and marital satisfaction in spouses of Operation Iraqi Freedom veterans: Relationships with spouses' perceptions of veterans' experiences and symptoms. *Journal of Family Psychology, 22*(4), 586–594. https://doi.org/10.1037/0893-3200.22.3.586

Rodriguez, K. E., Anderson, L. M., Ott, C. A., & O'Haire, M. E. (2021). The effect of a PTSD service dog on military veterans' medication regimens: A cross-sectional pilot study. *Anthrozoos, 34*(3), 393–406. https://doi.org/10.1080/08927936.2021.1898219

Rodriguez, K. E., Bryce, C. I., Granger, D. A., & O'Haire, M. E. (2018). The effect of a service dog on salivary cortisol awakening response in a military population with posttraumatic stress disorder (PTSD). *Psychoneuroendocrinology, 98*, 202–210. https://doi.org/10.1016/j.psyneuen.2018.04.026

Rodriguez, K. E., LaFollette, M. R., Hediger, K., Ogata, N., & O'Haire, M. E. (2020). Defining the PTSD service dog intervention: Perceived importance, usage, and symptom specificity of psychiatric service dogs for military veterans. *Frontiers in Psychology, 11*. https://doi.org/10.3389/fpsyg.2020.01638

Scotland-Coogan, D. (2019). Relationships, socialization and combat veterans: The impact of receiving and training a service dog. *Qualitative Report, 24*(8), 1897–1914. https://doi.org/10.46743/2160-3715/2019.3590

Scotland-Coogan, D., Whitworth, J. D., & Wharton, T. (2020). Outcomes of participation in a service dog training program for veterans with PTSD. *Society and Animals, 141*(1), 1–22. https://doi.org/10.1163/15685306-00001682

U.S. Department of Veterans Affairs. (2020). *A randomized trial of differential effectiveness of service dog pairing versus emotional support dog pairing to improve quality of life for veterans with PTSD*. www.research.va.gov/REPORT-Study-of-Costs-and-Benefits-Associated-with-the-Use-of-Service-Dogs-Monograph1.pdf.

Vitztum, C., & Urbanik, J. (2016). Assessing the dog: A theoretical analysis of the companion animal's actions in human-animal interactions. *Society and Animals, 24*(2), 172–185. https://doi.org/10.1163/15685306-12341399

Weathers, F., Litz, B., Herman, D., Huska, J., & Keane, T. (1993). The PTSD Checklist (PCL): Reliability, validity, and diagnostic utility. In *Paper presented at the annual convention of the international society for traumatic stress studies*.

Whitworth, J., O'Brien, C., Wharton, T., & Scotland-Coogan, D. (2020). Understanding partner perceptions of a service dog training program for veterans with PTSD: Building a bridge to trauma resiliency. *Social Work in Mental Health, 18*(6), 604–622. https://doi.org/10.1080/15332985.2020.1806181

Whitworth, J. D., Scotland-Coogan, D., & Wharton, T. (2019). Service dog training programs for veterans with PTSD: Results of a pilot controlled study. *Social Work in Health Care, 58*(4), 412–430. https://doi.org/10.1080/00981389.2019.1580238

Yarborough, B. J. H., Stumbo, S. P., Yarborough, M. T., Owen-Smith, A., & Green, C. A. (2018). Benefits and challenges of using service dogs for veterans with posttraumatic stress disorder. *Psychiatric Rehabilitation Journal, 41*(2), 118–124. https://doi.org/10.1037/prj0000294

Yount, R. A., Olmert, M. D., & Lee, M. R. (2012). Service dog training program for treatment of posttraumatic stress in service members. *United States Army Medical Department Journal*, 63–69.

Zilcha-Mano, S., Mikulincer, M., & Shaver, P. R. (2011). Pet in the therapy room: An attachment perspective on Animal-Assisted Therapy. *Attachment and Human Development, 13*(6), 541–561. https://doi.org/10.1080/14616734.2011.608987

CHAPTER 6

Equine-assisted therapy for veterans with PTSD

Diane Scotland-Coogan
School of Education and Social Services, Saint Leo University, Saint Leo, FL, United States

Introduction

Since September 11, 2001, the United States has deployed approximately 2.77 million service members to combat zones as a part of Operation Enduring Freedom, Operation Iraqi Freedom, and Operation New Dawn. Research shows those deployed to combat zones experience high rates of posttraumatic stress disorder (PTSD, 5.6%–30.5%) and traumatic brain injury (TBI, 10%–20%). Disorders highly comorbid with PTSD include depression and alcohol abuse, which may negatively impact veterans' cognitive and emotional functioning (Kinney et al., 2019; Norman et al., 2018). Adding the potential physical injuries incurred during military service, which may cause persistent pain, one can understand why veterans may experience significant barriers to successful reintegration to civilian life, impaired social functioning, and the development of much-needed support systems (Hoge et al., 2004; Kinney et al., 2019). Since 2008, suicides in the veteran population have increased, exceeding 6300. "The average number of Veteran suicide deaths per day has remained between 17 and 18, despite observed decreases in the size of the Veteran population" (National veteran suicide prevention annual report, 2020). More than 50% of all military members and veterans with deployment-related trauma issues resist participating in more traditional treatments, such as prolonged exposure therapy and cognitive processing therapy (Hoge et al., 2014; Monson et al., 2006; Scotland-Coogan et al., 2020). Other reports of veterans' participation in these programs state a 50% early dropout rate (Hoge et al., 2014; Scotland-Coogan et al., 2020). Reported reasons for this resistance to participate and early termination of treatment include time limitations, client hesitancy to talk about their trauma experiences, disconnects with the therapist, a perceived stigma of asking for help, and a desire not to be seen as "weak" when acknowledging their symptoms (Hoge et al., 2014; Scotland-Coogan et al., 2020). Even for those who participate in established

Animal Assisted Therapy Use Application by Condition
ISBN 978-0-323-98815-5
https://doi.org/10.1016/B978-0-323-98815-5.00001-X

treatment methods such as exposure-based therapies, a significant number will still meet the diagnostic criteria for PTSD (Kinney et al., 2019). Treatments that address these barriers are needed to work with our warriors.

Military culture and treatment

The veteran population has a very distinct culture. If one is not familiar with this culture, it can make working with them less than optimal. Therefore, understanding military culture becomes central to working with veterans (Coll et al., 2011; Understanding the military: The institution, the culture and the people, 2010). Veterans' trust in military comrades does not carry over to civilians; thus, civilian mental health providers may have difficulty establishing a therapeutic alliance. One should, therefore, consider a strength-based approach that focuses on what they can do rather than what they can no longer do. Treatment should include regaining a sense of self-mastery, which is an understanding that one has some control over their world (Naste et al., 2018). Self-mastery will help veterans see themselves as capable of making change and taking charge of their healing. Self-mastery is especially important if the veteran is struggling with symptoms of PTSD.

Posttraumatic stress disorder

Diagnostic criteria

The American Psychiatric Association Diagnostic and Statistical Manual of Mental Disorders, fifth ed. (Diagnostic and statistical manual of mental disorders, 2013) has specific criteria for diagnosing PTSD. These include Criterion A, "exposure to actual or threatened death, serious injury, or sexual violence." This exposure includes directly experiencing or witnessing the event in person as it happened to others, learning about a violent or accidental traumatic event of a family member or friend, or "experiencing repeated or extreme exposure to aversive details of the traumatic event(s)" (p. 271). Criterion B requires the presence of intrusive symptoms related to the trauma, "Recurrent, involuntary, and intrusive distressing memories," dreams or nightmares of the traumatic event(s), flashbacks, and intense reactions to cues related to the trauma (p. 271). Criterion C requires avoidance of any stimuli related to the trauma, such as memories, thoughts, or feelings, as well as external reminders, such as people and places (p. 271). Criterion D notes "negative alterations in cognitions and mood associated with the traumatic event(s)" with gaps in the information surrounding the traumatic event, and negative self-messages and perceptions of their world, diminished interest in participating in any activities, detachment from others, and the "inability to experience positive emotions" (p. 271–272). Criterion E requires altered reactions, including anger and "verbal or physical aggression toward people or objects," risky behaviors, hypervigilance, exaggerated startle experience, lack of focus and concentration, sleep disturbance, all of which have been experienced for more than one month (p. 272). Criterion F requires a duration of more than one month (p. 272). Criterion G requires these disturbances cause "clinically significant distress or impairment in social, occupational, or other important areas of functioning" (p. 272). Criterion H notes these disturbances cannot be the result of substance use or a medical condition (p. 272).

Symptoms of posttraumatic stress disorder

Symptoms of PTSD include anger, anxiety, sleep disturbances, poor self-image and perception of self-mastery, impaired executive function, isolation, problematic relationships, behavioral problems, substance abuse, depression, impaired focus and concentration, and suicide ideation, all of which are complexly interdependent (Olff et al., 2014; Simons et al., 2018; Twamley et al., 2009; Walter et al., 2010; Wirth, 2015). The individual will experience good and bad days, and remission of symptoms followed by more challenging times (Simons et al., 2018). The impact of the symptoms of PTSD is persistent and pervasive for veterans and their families. Psychoeducation about the symptoms and their effect is an essential piece for the veteran and their family.

Support of family and friends is essential to treatment adherence, rehabilitation, and overcoming the stigma surrounding mental health in the military (Meis et al., 2010; Schumm et al., 2013; Dekel & Monson, 2010). Walking on eggshells, loved ones consistently alter their behaviors to avoid triggering a negative reaction (Mansfield et al., 2014). Many spouses will choose to leave; others stay due to a sense of duty or a fear of their loved one committing suicide, suggesting that including the family in treatment should be considered (Dekel et al., 2005; Mansfield et al., 2014). Peer support, especially therapy with other veterans with similar experiences, has been shown to be valuable (Burton et al., 2019). This support may replace the comrade they lost when leaving the military.

Depression is highly correlated with PTSD. Suicide and suicide rates are twice as high in the veteran population (Brooks Holliday et al., 2016; Schaller et al., 2014). Substance abuse is prevalent and also highly correlated with PTSD. Self-medicating to manage symptoms of PTSD may drive the comorbidity of this disorder with substance abuse (Brown & Wolfe, 1994; Seal et al., 2011; Tipps et al., 2014). Domestic violence is more prevalent, which may be attributed to the anger and anxiety experienced with PTSD. Research has shown an incident rate of domestic violence five times higher in the military community (National Council of Child Abuse and Family Violence, 2015; Sherman et al., 2006; Tucker, 2009). This higher rate of domestic violence may, in part, be attributed to the overactive stress response causing symptoms of PTSD, which impairs executive function and promotes intolerance and anger.

Physiological and psychological symptoms of this disorder make daily functioning a struggle. Physiological effects of PTSD are a result of a stress response that no longer calms itself, manifesting in problems with anger, frustration, irritability, sleep, focus, exaggerated startle response, and hypervigilance (Olff et al., 2014; Twamley et al., 2009; Walter et al., 2010; Whealin, 2015). The effects of stress may either enhance or impair cognitive function. With a normal stress response, enhancement will occur, leaving one with a strong memory of a stressful event. However, when stress is persistent, such as with PTSD, cognitive function is impaired. These impairments impact intellectual functioning, executive functioning, attention/working memory, decision-making, reasoning, verbal memory, processing speed, and learning (Olff et al., 2014; Twamley et al., 2009; Walter et al., 2010; Wirth, 2015). Executive functioning regulates mood and impulse control, but when impaired, one may find it challenging to maintain relationships, manage self-care, and work independently (Walter et al., 2010). This impairment will impact treatment participation and success and potentially explain the high dropout rates of treatment-seeking veterans (Goetter et al., 2015; Hoge et al., 2014; Hudenko et al., 2016). Therefore, treatment should include addressing the physical nature of PTSD, the overactive stress response.

Approximately 50%–80% of veterans with PTSD also experience chronic pain and physical symptoms. This comorbidity makes treatment complicated, which may result in lower participation in treatment. Research indicates increases in physical symptoms coincided with increases in PTSD symptoms, but not the other way around; increases in PTSD did not necessarily predict an increase in physical symptoms (McAndrew et al., 2019). These findings would suggest an adjunct treatment to address physical symptoms.

Equine therapy treatment

Equine-assisted psychotherapy (EAP) is used as an adjunct therapy, combining psychotherapy modalities with equine therapy. The psychotherapy modalities may include cognitive behavioral therapy, mindfulness-based cognitive therapy, motivational interviewing, equine-facilitated cognitive processing therapy, and mindfulness-based stress reduction (Ferruolo, 2015; Wharton et al., 2019). These treatments are normally provided in isolation (i.e., without any other adjunct therapies simultaneously) (Kinney et al., 2019).

The equine-assisted therapy (EAT) team consists of an equine specialist, a therapist, and at least one horse (Fig. 6.1).

Figure 6.1 ***Equine therapy for veterans.*** Equine therapy animals with veterans and equine specialists. *(Photograph taken by Diane Scotland-Coogan. Written copyright permission on file.)*

While the equine specialist and therapist work closely together, they do not cross boundaries into each other's area of expertise (Parent, 2016). Equine specialists guide the therapist in choosing the appropriate horse. Because of their knowledge of the psychological and physical demands of the horse, the specialist will also select the location where therapy will take place. The equine specialist is also responsible for the safety of all involved in the session. They provide observations based on the horses' behaviors and, with the therapist, adhere to the treatment plan creating session follow-up activities that encourage and support the client in achieving their goals (Parent, 2016). The therapist's duties revolve around assisting individuals to improve clinical symptoms and to increase the veterans insight into their issues. Therapists will follow their chosen therapeutic modality to process the experiential activities performed with the horse(s). The therapist will pay particular attention to the clients' nonverbal communication during the activities and any metaphors used in speech. These are then addressed in the follow-up sessions (Parent, 2016). The therapist will suggest what processing activity may best support the client in achieving their goal. They may use empathic listening after the initial activity is set up and engage in debriefing after the session with the horse has ended (Fine, 2015). The therapist is responsible for the psychological safety of all (Parent, 2016).

Hippotherapy (HPOT) is an animal-assisted therapy with horses conducted by either an occupational therapist, speech therapist, or physical therapist. HPOT is a medical intervention typically used for those with physical disabilities. The American Hippotherapy Association (What is hippotherapy?, 2020) reports this form of EAT uses clinical reasoning and evidence-based practice "in the purposeful manipulation of equine movement as a therapy tool to engage sensory, neuromotor, and cognitive systems to promote functional outcomes (p. ?)." This therapy may help veterans with physical problems, including loss of limbs. Using the horse's natural gait, HPOT promotes physical improvement and healing (Fine, 2015). Most of the equine programs explored in this chapter use an on-the-ground (no riding) approach, including grooming the horse, leading the horse around an arena, as well as more difficult problem-solving tasks such as leading the horse through an obstacle course with no halter or lead line (Flora et al., 2016). These activities may assist in building self-mastery and self-confidence, as well as address physiological symptoms of PTSD.

EAT, EAP, or equine-facilitated psychotherapy are similar and are a form of experiential learning built around interactions with horses. For this chapter, EAT will be used to represent these forms of experiential learning.

Therapy is conducted by a provider who is trained and licensed by their state to practice psychotherapy or counseling. The therapist should have additional training and experience in facilitating EAT. Otherwise, sessions are facilitated by a trained therapist and an equine specialist, which is especially beneficial when working with a group. EAT will determine the client or group's treatment goals and identify ground activities that will help them meet these goals. Examples of skills applied in an EAT session include improved behaviors and social skills, depression and anxiety reduction, relationship development, and a more positive appraisal of self (Kinney et al., 2019; Sievert, n.d.). Family communication and relationships between veterans with PTSD and their families can be quite a struggle. Vasher et al. (2017) studied the impact of EAT on veteran's family communication. This research indicated EAT positively impacted communication, interpersonal skills, and problem-solving within the family.

Horses "perceive, respond to, and learn from the impressions" to sensory stimuli they are exposed to, no matter how minor the stimuli may be. Horses' abilities of keen perception, cognition, memory, and emotions may derive from observing human body language, sensing changes in human biochemistry, or even vocal tones. Changes in their environment, relationships with other horses, and their physical health all contribute to how they will respond to stimuli (PATH International, 2022). Horses will mirror a reflection of self by their behaviors and emotions. If the veteran is in a state of anxiety or anger, the horse will respond accordingly, in a mirroring effect. If the veteran is distracted by daily issues from their personal life, the horse will ignore them. Eventually, the veteran may realize that when they take the lead and focus on the horse, the horse will pay attention (Gehrke et al., 2018). This dynamic may help the veteran realize how others perceive them and what changes they may need to make. Veterans come to sessions with established concepts and opinions which are not serving them in their social environment. A plan is put in place by the team. The team puts together a plan for how to proceed. The experiential activity with the horse then provides immediate feedback through the horse's responses as to the success of the veteran's behaviors. This experience is then reflected on and processed to provide insight into what happened. The equine specialist provides observations of the horse's behaviors. When processed with the therapist, generalizations and concepts about the experience may lead to better understanding and insight into their symptoms. This newfound knowledge may then be used in other life situations (Parent, 2016). Changes made may improve relationships, provide an enhanced support system, assist in recognizing when they are experiencing symptoms of anxiety and calm themselves, and identify foundations of anger. This process is a form of experiential learning.

Experiential learning

Experiential educators include mental health professionals who engage with their clients in actual experiences, followed by reflection to increase awareness and knowledge, develop skills, and understand behaviors. The Association for Experiential Education (Association for Experiential Education, 2022) identifies experiential learning as:

- Experiential learning occurs when carefully chosen experiences are supported by reflection, critical analysis, and synthesis.
- Experiences are structured to require the learner to take initiative, make decisions, and be accountable for results.
- Throughout the experiential learning process, the learner is actively engaged in posing questions, investigating, experimenting, being curious, solving problems, assuming responsibility, being creative, and constructing meaning.
- Learners are engaged intellectually, emotionally, socially, soulfully, and/or physically. This involvement produces a perception that the learning task is authentic.
- The results of the learning are personal and form the basis for future experience and learning.
- Relationships are developed and nurtured: learner to self, learner to others, and learner to the world at large.
- The educator and learner may experience success, failure, adventure, risk-taking, and uncertainty because the outcomes of experience cannot totally be predicted.
- Opportunities are nurtured for learners and educators to explore and examine their own values.
- The educator's primary roles include setting suitable experiences, posing problems, setting boundaries, supporting learners, ensuring physical and emotional safety, and facilitating the learning process.
- The educator recognizes and encourages spontaneous opportunities for learning.
- Educators strive to be aware of their biases, judgments, and preconceptions and how these influence the learner.

- The design of the learning experience includes the possibility to learn from natural consequences, mistakes, and successes.

Experiential learning may have more of an impact on the veterans as it is a strength-based form of learning.

Ethical practices

Ethical practices of EAT address both the client and the horse. For clients, the Equine-Assisted Growth and Learning Association (Equine Assisted Growth and Learning Association, 2020) notes in their Code of Ethics, the commitment is to provide the highest quality of care to support clients in services in learning and growth. The provider will, in the best interest of the client, respect the value and dignity, as well as protect the safety and welfare of all clients. Referrals will be made in the event of lack of progress, and the provider will evaluate their own professional strengths and limitations in working with clients (Equine Assisted Growth and Learning Association, 2020).

With horses, ethical practices are used in both the caring for and working with the horses. The American Horse Council's National Welfare Code of Practice (American Horse Council, 2022) requires a commitment to "the dignity, humane care, health, safety, and welfare of horses in all our activities and care." This commitment includes all aspects of horse care, such as using appropriate training methods, respecting the horse's abilities and limits, the welfare of the horse taking precedence over the expectations of the therapeutic team. Facilities should be safe and appropriate, and the dignity, humane care, welfare, safety, and health of horses in all activities, ensuring that horses will be transitioned to more appropriate activities or uses as a need arises (American Horse Council, 2022).

Equine therapy with veterans

PTSD can be very complicated to treat. Veterans experience physical and environmental problems; these problems may be associated with mental health, substance abuse, and homelessness, all of which may all be linked to trauma exposure (Davis-Berman et al., 2018). Veterans are at significant risk for various mental health concerns, including depression, self-regulation issues, interpersonal issues, and suicide ideation and gestures (Wagenfeld et al., 2013). Ineffective treatment is linked to three-times higher suicide risk. Traditional approaches may not be effective. Additionally, there are barriers to treatment, such as the stigma associated with pursuing mental health treatment in the military culture (Davis-Berman et al., 2018). Individual therapy focuses on issues such as respect, boundaries, self-concept, appropriate assertiveness, and trust. Couples, family, or group therapy addresses communication, problem-solving, cooperation, and creative thinking. In these sessions, clients must be able to interact with each other as well as the horses (Masini, 2010). Adjunct treatments which are more acceptable to the veteran, such as animal-assisted therapy, should be included with established treatment approaches for optimum success. Monroe et al. (Monroe et al., 2019) noted when comparing EAT to traditional therapy-based approaches with veterans, only 12% discontinued participation, compared to an almost 50% dropout rate with the more traditional approaches (Hoge et al., 2014; Scotland-Coogan et al., 2020).

Family considerations

Deployment can cause communication issues as the family must find ways to cope emotionally and physically while their loved one is gone. The family becomes self-sufficient during the deployment and then must alter their new roles and routines upon the veteran's return. Rabenhorst et al. (2012), in an 85-month study analyzing substantiated spousal abuse reports for all married active-duty Air Force personnel noted increases in the rates of spousal abuse comparing pre- and post-deployment. The support of a spouse is crucial to treatment adherence and progress (Meis et al., 2010; Schumm et al., 2013), as well as to diminish the impact of the stigma of receiving mental health services (Dekel & Monson, 2010). If PTSD is left untreated, often the spouse will choose to leave (Dekel et al., 2005; Mansfield et al., 2014). Those who stay may suffer overwhelming loneliness and experience feelings of disconnect and neglect (Beks, 2016). Equine-assisted interventions have been used in couple's therapy and found to be beneficial in treating trauma-related disorders in veterans and their spouses (Willmund et al., 2021). Flora et al. (2016) suggest that multiple sessions of EAT can help families reconfigure their story, which is currently filled with problem-saturated narratives. The family is assisted in creating a new, collaborative family story to promote family reintegration and to improve communication (Monroe et al., 2019).

The Professional Association of Therapeutic Horsemanship, Int. (PATH) (PATH International, 2022) states veterans with physiological dysregulation and mobility impairments benefit through the horses' gait, which is similar to humans. Improvements can be achieved in balance, flexibility, posture, core strength, and self-confidence. Socially, veterans can feel a loss of camaraderie after leaving the service. Group EAT provides them with the opportunity to recreate their military community through other veterans, staff, and horses. Reducing isolation and the feeling of disconnect from others can be healing for veterans (PATH International, 2022). Cognitively, the structure required to participate in EAT is like the structure they experienced in the military. Executive functioning, including focus, reasoning, organization, and attention, is required to complete EAT activities. Emotionally, the veteran bonds with the horse, which can be comforting and address avoidant attachment, thus improving interpersonal relationships (Zilcha-Mano et al., 2011). Self-confidence and self-worth can be developed by practicing skills such as emotional regulation, distress tolerance, and developing resilience, all contributing to the healing of PTSD (PATH International, 2022). Veterans may find it easier to explore their own transference and projection responses through working with a horse. The horse's hypervigilance and acute awareness of a perceived threat are similar to the veteran with PTSD, making it easier for the veteran to relate to the horse (Monroe et al., 2019).

Veterans are not always aware of how others perceive them; having the horse respond to them may assist them in developing some insight as the horse mirrors back their perceptions of the veterans' actions. Veterans may develop a sense of self-mastery as they learn to communicate nonverbally with the horse. Working with a horse requires the veteran to learn to tune into their emotions, manage them, and develop new ways of thinking and responding. Horses are nonjudgmental mirrors, which requires the veteran to be aware of how they are internalizing their thoughts and emotions and how others perceive them. Horses are prey animals; they have had to be constantly aware of their surroundings to keep themselves safe for survival. They can be relaxed, yet ready, which can assist veterans in EAT to develop the ability to be relaxed yet ready through awareness of their emotions, eventually developing the ability to self-regulate (Trotter, 2012). Therapy sessions should repeat weekly to allow the veteran to explore and process emerging thoughts and feelings with the therapist (Fine, 2015). Berget et al. (Berget et al., 2008) found that participants continued to demonstrate significantly improved self-efficacy when using animal-assisted therapy with farm animals at the six-month follow-up period.

Lac (2020) noted the following themes which may emerge in EAT as difficult (p. 92):

- Trusting themselves to know what is safe or not
- Recognizing boundaries
- Challenges and/or benefits in staying in the present moment
- What is a safe connection (with horses and humans)?
- Identifying issues around choice/freedom
- Authenticity
- Shame and vulnerability

Substance abuse disorder and equine therapy

The prevalence of PTSD among veterans with alcohol use disorder (AUD) was 20.3%. PTSD was more than four times more prevalent among veterans with AUD than observed in the general US adult population (Norman et al., 2018). Staying in and completing treatment for AUD is known to support improved outcomes. The dropout rate for substance abuse treatment often exceed 50% (Kern-Godal, Brenna, Arnevik, et al., 2016). Participants tend to remain in treatment longer and be more apt to complete treatment in an equine-assisted substance abuse treatment program. The relationship with the horse assisted in the veteran feeling understood and accepted (Kern-Godal, Brenna, Kogstad, et al., 2016).

Kern-Godal, Brenna, Arnevik, et al.(2016) found the stable seemed to provide a safe space in which participants could create a positive sense of self, one which differs from the patient role. Instead, veterans see themselves as competent within this setting rather than one who is broken. They feel their presence is necessary; they are accepted and can deal with challenges, leading to improved self-mastery. Participants looked forward to their time at the stable, which is in contrast to many more traditional treatments, which may explain the low EAT dropout rates.

What does this mean for veterans?

Given all the information on adjunct treatments for PTSD, one needs to stop evaluating the needs of our combat veterans through modalities developed more than 50 years ago (Gillett & Weldrick, 2014). There are many factors contributing to the development and maintenance of mental health. Examining PTSD through a physiological, psychological, and spiritual lens suggests one should seek to treat mental health disorders in a way that addresses all these factors.

Independence and self-mastery

Gehrke et al. (2018) noted a 20% increase in self-perceived quality of life after veterans participated in a four-week EAT program. EAT has demonstrated increased self-control, self-image, personal growth, self-awareness, self-esteem, and life satisfaction in veterans with PTSD, reporting improvements in patience, trust, the ability to relax, and coping skills to manage anger, stress, and frustration (Romaniuk et al., 2018). An increase in the ability to tolerate changes in their environment was also reported. Veterans reported improved quality of life, based on all of their growth in EAT (Burton et al., 2019). Veterans have reported such benefits as learning about themselves, respect, and spiritual connection. In addition, the mirroring effect, the ability to mirror back human emotions, is the foundation of the horse/human connection. This connection may also promote a sense of self-efficacy, which builds self-confidence, all of which are problematic for a veteran with PTSD (Sylvia et al., 2020).

Focus, concentration, and memory

Declarative memory consists of memory for facts and events (Riedel & Blokland, 2015). Research has indicated declarative memory impairment in combat veterans with PTSD over combat veterans without PTSD (Burriss et al., 2008). The Wechsler Adult Intelligence Scale-IV Verbal Comprehension Index measures one's ability to use, understand, and think with spoken language. It can also reflect the depth of knowledge acquired within one's environment and the ability to retrieve information from long-term memory (Greenwald, 2015). Combat veterans with PTSD, when compared to a similar group of combat veterans without PTSD, have demonstrated decreased scores on the Verbal Comprehension Index. They have also exhibited impairment in learning and memory tasks (Burriss et al., 2008), suggesting difficulty in participating in talk therapy, diminished ability to communicate their thoughts, and inability to follow through with tasks at the very least. The use of EAT to lessen the symptoms of PTSD contributing to these deficits without relying on words or memory may assist in improving focus, concentration, and memory.

Relationships

Communication and relationships between veterans with PTSD and their families can be quite a struggle. Vasher et al. (2017) reviewed the impact of EAT on veterans' family communication and found EAT positively impacted communication, interpersonal skills, and problem-solving within the family. EAT also led to improvements in communication with the outside world (Burton et al., 2019). Other enhanced areas included patience and trust (Burton et al., 2019; Romaniuk et al., 2018). This change may be attributed to working together as a team, being more openly communicative with each other in sessions, and the improvement of PTSD symptoms noted with equine therapy.

Anxiety, depression, and anger

EAT has been identified as for those who may struggle with depression, anger, fear, and anxiety. Breaking through one's defensive barriers, EAT allows the developing of new insight and perspectives from old behavior patterns and relationships (Burriss et al., 2008; Burton et al., 2019). Horses allow veterans to be more open to receiving criticism and manage self-judgment. This openness can assist in dealing with emotional upheavals, stress, and management of anxiety symptoms. The confidence developed may support more confidence in their decision-making abilities (Gehrke et al., 2018). Fisher et al. (2021) found that not only did EAT reduce symptoms of PTSD, but also the effect was still in place three months later.

Veterans continue to seek alternative ways to manage their symptoms of PTSD. Treatment for PTSD should first address the benefits of calming the hyperarousal response and then address psychological symptoms when the veteran is more able to participate in a treatment requiring cognitive functioning. Also, one should consider adjunct treatment, which does not carry the stigma of more traditional forms of therapy. EAT has shown promise in treating anxiety, depression, elevating self-confidence, self-concept, self-esteem, resulting in an overall improvement in well-being, all of which assist this population with reintegration into civilian life (Ferruolo, 2015).

References

American Horse Council. (2022). *National welfare code of practice*. www.horsecouncil.org/resources/national-welfare-code-2/

Association for Experiential Education. (2022). *What is experiential education?*. www.aee.org/what-is-experiential-education

Beks, T. A. (2016). Walking on eggshells: The lived experience of partners of veterans with PTSD. *Qualitative Report, 21*(4), 645–660. http://nsuworks.nova.edu/cgi/viewcontent.cgi?article=2269&;context=tqr

Berget, B., Ekeberg, Ø., & Braastad, B. O. (2008). Animal-assisted therapy with farm animals for persons with psychiatric disorders: Effects on self-efficacy, coping ability and quality of life, a randomized controlled trial. *Clinical Practice and Epidemiology in Mental Health, 4*. https://doi.org/10.1186/1745-0179-4-9

Brooks Holliday, S., Pedersen, E. R., & Leventhal, A. M. (2016). Depression, posttraumatic stress, and alcohol misuse in young adult veterans: The transdiagnostic role of distress tolerance. *Drug and Alcohol Dependence, 161*, 348–355. https://doi.org/10.1016/j.drugalcdep.2016.02.030

Brown, P. J., & Wolfe, J. (1994). Substance abuse and post-traumatic stress disorder comorbidity. *Drug and Alcohol Dependence, 35*(1), 51–59. https://doi.org/10.1016/0376-8716(94)90110-4

Burriss, L., Ayers, E., Ginsberg, J., & Powell, D. A. (2008). Learning and memory impairment in PTSD: Relationship to depression. *Depression and Anxiety, 25*(2), 149–157. https://doi.org/10.1002/da.20291

Burton, L. E., Qeadan, F., & Burge, M. R. (2019). Efficacy of equine-assisted psychotherapy in veterans with posttraumatic stress disorder. *Journal of Integrative Medicine, 17*(1), 14–19. https://doi.org/10.1016/j.joim.2018.11.001

Coll, J. E., Weiss, E. L., & Yarvis, J. S. (2011). No one leaves unchanged: Insights for civilian mental health care professionals into the military experience and culture. *Social Work in Health Care, 50*(7), 487–500. https://doi.org/10.1080/00981389.2010.528727

Davis-Berman, J., Berman, D., & Berman, N. D. (2018). Outdoor programs as treatment for posttraumatic stress disorder in veterans: Issues and evidence. *Best Practices in Mental Health, 14*(2), 9–20.

Dekel, R., Goldblatt, H., Keidar, M., Solomon, Z., & Polliack, M. (2005). Being a wife of a veteran with posttraumatic stress disorder. *Family Relations, 54*(1), 24–36. https://doi.org/10.1111/j.0197-6664.2005.00003.x

Dekel, R., & Monson, C. M. (2010). Military-related post-traumatic stress disorder and family relations: Current knowledge and future directions. *Aggression and Violent Behavior, 15*(4), 303–309. https://doi.org/10.1016/j.avb.2010.03.001

Diagnostic and statistical manual of mental disorders. (2013). American Psychiatric Association. https://doi.org/10.1176/appi.books.9780890425596

Equine Assisted Growth and Learning Association. (2020). *EGALA code of ethics*. www.eagala.org/ethics

Ferruolo, D. M. (2015). Psychosocial equine program for veterans. *Social Work, 61*(1), 53–60. https://doi.org/10.1093/sw/swv054

Fine, A. (Ed.). (2015). *Handbook on animal-assisted therapy: Foundations and guidelines for animal-assisted interventions*. Elsevier Science & Technology.

Fisher, P. W., Lazarov, A., Lowell, A., Arnon, S., Turner, J. B., Bergman, M., Ryba, M., Such, S., Marohasy, C., Zhu, X., Suarez-Jimenez, B., Markowitz, J. C., & Neria, Y. (2021). Equine-assisted therapy for posttraumatic stress disorder among military veterans. *The Journal of Clinical Psychiatry, 82*(5). https://doi.org/10.4088/jcp.21m14005

Flora, J., Boje, D., Rosile, G. A., & Hacker, K. (2016). A theoretical and applied review of embodied restorying for post-deployment family reintegration. *Journal of Veterans Studies, 1*(1), 129. https://doi.org/10.21061/jvs.41

Gehrke, E. K., Noquez, A. E., Ranke, P. L., & Myers, M. P. (2018). Measuring the psychophysiological changes in combat Veterans participating in an equine therapy program. *Journal of Military, Veteran and Family Health, 4*(1), 60–69. https://doi.org/10.3138/jmvfh.2017-0015

Gillett, J., & Weldrick, R. (2014). *Effectiveness of psychiatric service dogs in the treatment of post-traumatic stress disorder among veterans*. McMaster University.

Goetter, E. M., Bui, E., Ojserkis, R. A., Zakarian, R. J., Brendel, R. W., & Simon, N. M. (2015). A systematic review of dropout from psychotherapy for posttraumatic stress disorder among Iraq and Afghanistan combat veterans. *Journal of Traumatic Stress, 28*(5), 401–409. https://doi.org/10.1002/jts.22038

Greenwald, V. (2015). *Wechsler adult intelligence scale-IV: Description of WAIS-IV indexes*. http://washingtoncenterforcognitivetherapy.com/wp-content/uploads/2015/01/greenwood_description-wais-1.pdf#:~:text=The%20vocabulary%20subtest%20requires%20the%20client%20to%20try,appropriately%20retrieval%20of%20information%20-from%20long-term%20memory%20Information. http://%20

Hoge, C. W., Castro, C. A., Messer, S. C., McGurk, D., Cotting, D. I., & Koffman, R. L. (2004). Combat duty in Iraq and Afghanistan, mental health problems, and barriers to care. *New England Journal of Medicine, 351*(1), 13–22. https://doi.org/10.1056/NEJMoa040603

Hoge, C. W., Grossman, S. H., Auchterlonie, J. L., Riviere, L. A., Milliken, C. S., & Wilk, J. E. (2014). PTSD treatment for soldiers after combat deployment: Low utilization of mental health care and reasons for dropout. *Psychiatric Services, 65*(8), 997–1004. https://doi.org/10.1176/appi.ps.201300307

Hudenko, W., Homaifar, B., & Wortzel, H. (2016). The relationship between PTSD and suicide. In *National center for posttraumatic stress disorders*. Department of Veterans Affairs.

Kern-Godal, A., Brenna, I. H., Arnevik, E. A., & Ravndal, E. (2016). More than just a break from treatment: How substance use disorder patients experience the stable environment in horse-assisted therapy. *Substance Abuse: Research and Treatment, 10*, 99–108. https://doi.org/10.4137/SART.S40475

Kern-Godal, A., Brenna, I. H., Kogstad, N., Arnevik, E. A., & Ravndal, E. (2016). Contribution of the patient-horse relationship to substance use disorder treatment: Patients' experiences. *International Journal of Qualitative Studies on Health and Well-Being, 11*. https://doi.org/10.3402/qhw.v11.31636

Kinney, A. R., Eakman, A. M., Lassell, R., & Wood, W. (2019). Equine-assisted interventions for veterans with service-related health conditions: A systematic mapping review. *Military Medical Research, 6*(1). https://doi.org/10.1186/s40779-019-0217-6

Lac, V. (2020). *It's not about the activity: Thinking outside the toolbox in equine facilitated psychotherapy and learning*. University Professors Press.

Mansfield, A. J., Schaper, K. M., Yanagida, A. M., & Rosen, C. S. (2014). One day at a time: The experiences of partners of veterans with posttraumatic stress disorder. *Professional Psychology: Research and Practice, 45*(6), 488–495. https://doi.org/10.1037/a0038422

Masini, A. (2010). Equine-assisted psychotherapy in clinical practice. *Journal of Psychosocial Nursing and Mental Health Services, 48*(10), 30–34. https://doi.org/10.3928/02793695-20100831-08

McAndrew, L. M., Lu, S. E., Phillips, L. A., Maestro, K., & Quigley, K. S. (2019). Mutual maintenance of PTSD and physical symptoms for Veterans returning from deployment. *European Journal of Psychotraumatology, 10*(1). https://doi.org/10.1080/20008198.2019.1608717

Meis, L. A., Barry, R. A., Kehle, S. M., Erbes, C. R., & Polusny, M. A. (2010). Relationship adjustment, PTSD symptoms, and treatment utilization among coupled national guard soldiers deployed to Iraq. *Journal of Family Psychology, 24*(5), 560–567. https://doi.org/10.1037/a0020925

Monroe, M., Whitworth, J. D., Wharton, T., & Turner, J. (2019). Effects of an equine-assisted therapy program for military veterans with self-reported PTSD. *Society & Animals, 29*(5–6), 577–590. https://doi.org/10.1163/15685306-12341572
Monson, C. M., Schnurr, P. P., Resick, P. A., Friedman, M. J., Young-Xu, Y., & Stevens, S. P. (2006). Cognitive processing therapy for veterans with military-related posttraumatic stress disorder. *Journal of Consulting and Clinical Psychology, 74*(5), 898–907. https://doi.org/10.1037/0022-006X.74.5.898
Naste, T. M., Price, M., Karol, J., Martin, L., Murphy, K., Miguel, J., & Spinazzola, J. (2018). Equine facilitated therapy for complex trauma (EFT-CT). *Journal of Child and Adolescent Trauma, 11*(3), 289–303. https://doi.org/10.1007/s40653-017-0187-3
National Council of Child Abuse and Family Violence. (2015). *Combat vets most prone to domestic abuse.* http://www.nccafv.org/combat_vets.htm
National veteran suicide prevention annual report 2020.(2020). www.mentalhealth.va.gov/docs/data-sheets/2020/2020-National-Veteran-Suicide-Prevention-Annual-Report-11-2020-508.pdf
Norman, S. B., Haller, M., Hamblen, J. L., Southwick, S. M., & Pietrzak, R. H. (2018). The burden of co-occurring alcohol use disorder and PTSD in U.S. military veterans: Comorbidities, functioning, and suicidality. *Psychology of Addictive Behaviors, 32*(2), 224–229. https://doi.org/10.1037/adb0000348
Olff, M., Polak, A. R., Witteveen, A. B., & Denys, D. (2014). Executive function in posttraumatic stress disorder (PTSD) and the influence of comorbid depression. *Neurobiology of Learning and Memory, 112*, 114–121. https://doi.org/10.1016/j.nlm.2014.01.003
Parent, I. B. (2016). *Fundamentals of equine assisted trauma therapy.* CreateSpace Independent Publishing Platform
PATH International. (2022). *PATH international organization.* https://pathintl.org/
Rabenhorst, M. M., Thomsen, C. J., Milner, J. S., Foster, R. E., Linkh, D. J., & Copeland, C. W. (2012). Spouse abuse and combat-related deployments in active duty Air Force couples. *Psychology of Violence, 2*(3), 273–284. https://doi.org/10.1037/a0027094
Riedel, W. J., & Blokland, A. (2015). Declarative memory. *Handbook of Experimental Pharmacology, 228*, 215–236. https://doi.org/10.1007/978-3-319-16522-6_7
Romaniuk, M., Evans, J., Kidd, C., & Chao, L. (2018). Evaluation of an equine-assisted therapy program for veterans who identify as 'wounded, injured or ill' and their partners. *Plos one, 13*(9), e0203943. https://doi.org/10.1371/journal.pone.0203943
Schaller, E. K., Woodall, K. A., Lemus, H., Proctor, S. P., Russell, D. W., & Crum-Cianflone, N. F. (2014). A longitudinal comparison of posttraumatic stress disorder and depression among military service components. *Military Psychology, 26*(2), 77–87. https://doi.org/10.1037/mil0000034
Schumm, J. A., Fredman, S. J., Monson, C. M., & Chard, K. M. (2013). Cognitive-behavioral conjoint therapy for PTSD: Initial findings for operations enduring and Iraqi freedom male combat veterans and their partners. *American Journal of Family Therapy, 41*(4), 277–287. https://doi.org/10.1080/01926187.2012.701592
Scotland-Coogan, D., Whitworth, J. D., & Wharton, T. (2020). Outcomes of participation in a service dog training program for veterans with PTSD. *Society and Animals, 141*(1), 1–22. https://doi.org/10.1163/15685306-00001682
Seal, K. H., Cohen, G., Waldrop, A., Cohen, B. E., Maguen, S., & Ren, L. (2011). Substance use disorders in Iraq and Afghanistan veterans in VA healthcare, 2001–2010: Implications for screening, diagnosis and treatment. *Drug and Alcohol Dependence, 116*(1–3), 93–101. https://doi.org/10.1016/j.drugalcdep.2010.11.027
Sherman, M. D., Sautter, F., Jackson, M. H., Lyons, J. A., & Han, X. (2006). Domestic violence in veterans with posttraumatic stress disorder who seek couples therapy. *Journal*

of Marital and Family Therapy, 32(4), 479–490. https://doi.org/10.1111/j.1752-0606.2006.tb01622.x

Sievert, R. (n.d.). Equine assisted modalities: Glossary of terms. https://sdpsych.org/resources/Documents/Sievert_%20Equine%20Assisted%20Resources_Handout.pdf#:~:text=Renee%20Sievert%20Equus%20Coaching%20Equine%20Assisted%20Psychotherapy%20%28EAP%29%3AAddresses,defined%20by%20their%20treatment%20pl%20an%20or%20goals

Simons, J. S., Simons, R. M., Keith, J. A., Grimm, K. J., Stoltenberg, S. F., O'Brien, C., & Andal, K. (2018). PTSD symptoms and alcohol-related problems among veterans: Temporal associations and vulnerability. *Journal of Abnormal Psychology, 127*(8), 733–750. https://doi.org/10.1037/abn0000376

Sylvia, L., West, E., Blackburn, A. M., Gupta, C., Bui, E., Mahoney, T., Duncan, G., Wright, E. C., Lejeune, S., & Spencer, T. J. (2020). Acceptability of an adjunct equine-assisted activities and therapies program for veterans with posttraumatic stress disorder and/or traumatic brain injury. *Journal of Integrative Medicine, 18*(2), 169–173. https://doi.org/10.1016/j.joim.2020.01.005

Tipps, M. E., Raybuck, J. D., & Lattal, K. M. (2014). Substance abuse, memory, and post-traumatic stress disorder. *Neurobiology of Learning and Memory, 112*, 87–100. https://doi.org/10.1016/j.nlm.2013.12.002

Trotter, K. S. (2012). Harnessing the power of equine assisted counseling: Adding animal assisted therapy to your practice. In *Harnessing the power of equine assisted counseling: Adding animal assisted therapy to your practice* (pp. 1–309). Taylor and Francis. https://doi.org/10.4324/9780203802038

Tucker, D. D. (2009). *Domestic violence, PTSD and brain injury: Military and civilian challenges.* http://www.ncdsv.org/images/MFLJ_Domestic%20Violence%20and%20PTSD_Tucker_Pt%201_4-2009.pdf

Twamley, E. W., Allard, C. B., Thorp, S. R., Norman, S. B., Cissell, S. H., Berardi, K. H., Grimes, E. M., & Stein, M. B. (2009). Cognitive impairment and functioning in PTSD related to intimate partner violence. *Journal of the International Neuropsychological Society, 15*(6), 879–887. https://doi.org/10.1017/S135561770999049X

Understanding the military: The institution, the culture and the people. (2010). Substance Abuse and Mental Health Services Administration. www.samhsa.gov/sites/default/files/military_white_paper_final.pdf

Vasher, H., Delano, M., Herlache-Pretzer, E., Meyer, K., & Stawowy, K. (2017). The impact of equine-assisted therapy on communication within familial relationships of veterans with posttraumatic stress disorder. *The American Journal of Occupational Therapy, 71*(4_Suppl. ment_1). https://doi.org/10.5014/ajot.2017.71s1-po1021, 7111520272pp.1–7111520272p.1

Wagenfeld, A., Roy-Fisher, C., & Mitchell, C. (2013). Collaborative design: Outdoor environments for veterans with PTSD. *Facilities, 31*(9), 391–406. https://doi.org/10.1108/02632771311324954

Walter, K. H., Palmieri, P. A., & Gunstad, J. (2010). More than symptom reduction: Changes in executive function over the course of PTSD treatment. *Journal of Traumatic Stress, 23*(2), 292–295. https://doi.org/10.1002/jts.20506

Wharton, T., Whitworth, J., Macauley, E., & Malone, M. (2019). Pilot testing a manualized equine-facilitated cognitive processing therapy (EF-CPT) intervention for PTSD in veterans. *Psychiatric Rehabilitation Journal, 42*(3), 268–276. https://doi.org/10.1037/prj0000359

What is hippotherapy?. (2020). American Hippotherapy Association. www.americanhippotherapyassociation.org/what-is-hippotherapy

Whealin, J. M. (2015). *Warzone-related stress reactions: What veterans need to know*. Retrieved from http://www.ptsd.va.gov/professional/manuals/manual-pdf/iwcg/iraq_clinician_guide_app_j2.pdf

Willmund, G., Zimmermann, P., Alliger-Horn, C., Varn, A., Fischer, C., Parent, I., Sobottka, A., Bering, R., Rose, C., Ströhle, A., & Köhler, K. (2021). Equine-assisted psychotherapy with traumatized couples—improvement of relationship quality and psychological symptoms. *Journal of Marital and Family Therapy, 47*(4), 925–944. https://doi.org/10.1111/jmft.12485

Wirth, M. M. (2015). Hormones, stress, and cognition: The effects of glucocorticoids and oxytocin on memory. *Adaptive Human Behavior and Physiology, 1*(2), 177–201. https://doi.org/10.1007/s40750-014-0010-4

Zilcha-Mano, S., Mikulincer, M., & Shaver, P. R. (2011). An attachment perspective on human-pet relationships: Conceptualization and assessment of pet attachment orientations. *Journal of Research in Personality, 45*(4), 345–357. https://doi.org/10.1016/j.jrp.2011.04.001

CHAPTER 7

Avian-assisted therapy for posttraumatic stress disorder

Greg Para
Sarasota Parrot Conservatory, Myakka, FL, United States

Imagine gazing into a campfire and getting lost in your thoughts. Watching the flames dance around with wisps of blues, red, and orange flames licking the air. Bright yellow gusts of crackling fire sparkles in the night carrying you away with them like little fireflies flitting here and there. Most of us have felt the calmness that comes with sitting by a campfire and being present.

This is the closest analogy to the feeling one gets simply sitting and watching a parrot preen. Due to its nature, a parrot is constantly gathering oil from its gland by its tail and distributing it on each feather it has. This, when mixed with sunlight adds to the luxuriant nature of a parrot's feathers (Fig. 7.1).

It is vital to its health, but simply watching a parrot preen is vital to the health of someone suffering from posttraumatic stress disorder (PTSD).

Many combat veterans including myself have found that the simple act of watching a parrot preen clears our mind. One forgets about the roadside bombs, the snipers, the incoming mortars, and the hypervigilance. Bird watching has been a hobby for centuries and well documented for calming one's mind. Mounting evidence supporting the health benefits of the outdoors is helping shape innovations in medicine, education, and more (Adams, 2019).

When I describe to people what PTSD looks like, I show them a parrot that has suffered trauma and has plucked its feathers. Under severe stress and depression, a parrot will often pluck its feathers and in some instances self-mutilate; conditions well documented by avian veterinarians. Self-harm, including, unfortunately, suicide, can also occur in people who have suffered trauma. So, a similarity exists, and a bond can easily be obtained between people and parrots.

In his article "What does a parrot know about PTSD?" (Siebert, 2016) Charles Siebert wrote: "Abandoned pet parrots are twice-traumatized beings: denied first their natural will to flock and then the company of the humans who owned them. So, matching a human suffering from PTSD

Animal Assisted Therapy Use Application by Condition
ISBN 978-0-323-98815-5
https://doi.org/10.1016/B978-0-323-98815-5.00015-X

Figure 7.1 Man with parrot. Chapter author Greg Para with Forest, a Greenwing Macaw on his shoulder. *(Photo by Lori Ludwig, written copyright permission on file).*

with a parrot who suffers from PTSD can produce magical results as the parrot's natural inclination is to bond to be part of a flock again. Science has shown this innate desire is why parrots bond with humans."

Why parrots?

The number one question people ask me is, "can your parrot talk?" Most parrots do not talk, but many can; however, many can intuit your feelings and communicate with you which makes them an ideal companion. Parrots are very playful by nature, and many know how to be pets and have fun when you need it most. One of my parrots will know when I am down and will poke me in the neck to tickle me, and when I put my head close to stop her, she will poke me on the other side interacting with me. Sometimes she'll say, "give me a kiss," other times she will keep tickling me saying "stop," "stop it" knowing that's what I will say to her. Pretty soon I am engaged in interacting with her and notice my mood has picked up.

There are a number of studies and stories about the intelligence and capabilities of parrots. When it comes to parrot intelligence, the most famous parrot may be an African gray named Alex. Alex was the subject of a 30-year experiment (Pepperberg, 2008) by animal psychologist Irene Pepperberg, who worked at the University of Arizona and later at Harvard University and Brandeis University.

You will find that some parrots just repeat words, and they may or may not be timed appropriately for the situation. I had one parrot that would always say "I want to go out," "I want to go out." But even when I took him out, he would still say "I want to go out." I have another parrot that said, "you already are!" Now the cognitive ability of this parrot was instrumental when I returned from Afghanistan in healing. I spent many nights in her cage, and my story was documented in an episode of Animal Planet's "Collar of Duty" series (Collar of Duty: Greg & the Parrots: A U.S. war veteran diagnosed with PTSD finds solace in rescue parrots Bella and Chico, 2017).

I credit Bella and Chico, both blue and gold macaws with saving my life. Chico is a Chica, but oftentimes people will give a parrot a name without having it sexed. Let me back up a bit so I can take you through my personal journey healing from PTSD and helping others with PTSD heal through working with the parrots I have rescued at Sarasota Parrot Conservatory Inc.

A dog or a parrot

I believe that depending on the person and disability, no single animal is a panacea. Whereas dogs are faithful companions that can sleep with you, run and walk with you, and learn to chase a ball or intuit your mood, parrots can do the same, but an advantage is they can communicate verbally with you. In addition, they can eat the same meals out of the same plate or bowl with you, shower with you as most thoroughly enjoy getting misted, and will also intuit your moods and proactively work to improve it. Many people are surprised at the affection a parrot can offer. They will kiss, they will snuggle, they will become playful, tell you they love you, ask to step up, and lay on their backs to play like a puppy. Different animals have been used for assisted therapy and often it comes down to the choice of the person. Dogs have become the most popular and many have been trained to become service animals. Equine therapy is gaining popularity and has become an effective choice, but now more programs are beginning to offer therapy parrots through an adoption process. However, while you most likely need to own a dog as a service animal, you can also just visit a parrot rescue and enjoy the benefits of their therapeutic value without the responsibility of owning one. Owning a parrot requires a lot of work which is therapeutic, but a common description for owning a parrot is likened to having a perpetual two-year-old for life!

My story

In 2011, I was activated and deployed with the US Army to Afghanistan. I was a naval officer in the reserves but ended up going with the army.

I had a wife and four children as well as a dog, a cat, and a parrot. Bella was my first parrot, a blue and gold macaw. While deployed, whenever I got the chance, I would Skype home to speak with the family and Bella would recognize me on the screen and would say "hello," "goodbye," "night," and "I love you." Although I looked forward to talking with my family, I had a special need to communicate with Bella. I carried a dream catcher my wife had made for me with Bella's feathers, and I hung it in my tent. In many missions outside the wire as it is commonly called, I saw my share of frightening things and endured the constant feeling of hypervigilance waiting for the attacks which led to adrenal fatigue among other things. Upon my return from deployment, I was diagnosed with PTSD and also endured numerous surgeries for back and musculoskeletal injuries. Not everyone came back from missions, which is something many of us will never forget.

Bella was born with severe splay leg, so was unable to perch with her feet, walk or use her feet to eat. Imagine your ankles bent inward 90 degrees. Animals don't recognize disabilities, and they make do with what they have. She was an inspiration to me while I was deployed and certainly when I returned home. Trauma victims report they feel their life is out of control, chaotic, and unorganized. One of the many benefits of avian therapy is having the benefit of a routine. Parrots must have routines, and, when they don't, they become very upset. When working with a parrot, there is a therapeutic value to having a routine with your parrot. So, my healing began with a routine every day. I had to get up to feed Bella, then she would always want to interact with me, and during the day, she was always available; no questions asked when I needed to be with her. We also had our evening routine where just before dark I would get her and hold her on my chest while we sat on the couch. I would pet her, or she would turn onto her back, and I would tickle her belly and she would grab me with her beak like a puppy would and she would then tire of this banter and hunker down to rest. When parrots are content and before they fall asleep, they grind their beaks. This can be a very calming sound, and for me, it certainly was. She would softly caw and grind her beak, and I would just be still and soak this experience in. Finally, I would put her on the back of the couch where she would sleep. I still have this routine nightly with her going on 10 years.

Shortly after I returned from Afghanistan, I had the opportunity to rescue another blue and gold macaw and that is how Chico came into my life. Originally, she was found screaming "feed Chico," "feed Chico" in an apartment that had been vacated 10 days prior to her being found. Due to this stress of being abandoned, she plucked all her gold feathers on her breast and to this day only has patches of gray down feathers on her breast. Her feathers will never grow back. She changed my life!

Chico was a challenge at first, and it took me 110 days before she would allow me to hold her. She by far had the most intelligence of any parrot I have met. When I brought her home, I knew it was a strange place and I asked her if she was afraid, and she said "yes." I asked her if she wanted me to leave a light on in the aviary and again, she said "yes." Before I left, she said "give me a kiss," so I kissed her beak and then she said, "I love you." This was 10 years ago, and this routine still happens every evening before I go to bed. She whispers at night knowing you must be quiet. Many mornings the birds will start being noisy, and if Chico is not ready to get up, she will tell them to be quiet. She knows they are making noise, and she knows what to tell them. One morning I went to see Chico and her left wing kept drooping, and she would bite it and pull it back up. I asked her what was wrong with her wing. Her reply: "it hurts." I was floored. I wasn't sure I heard her correctly, so I asked her again what was wrong with her wing and again she answered, "it hurts." I brought her to the vet that day, and she must have had a scare during the night and hurt her wing flying in the aviary. Thanks to Bella (Fig. 7.2) and Chico (Fig. 7.3).

I began a healing journey that continues to this day. I recognized the healing qualities that parrots brought me, and I wanted to explore how I might

Figure 7.2 Bella, a blue and gold Macaw with Greg Para. Bella, a blue and gold Macaw, being held by the author wearing a jacket made in Afghanistan to match the color of Bella. *(Photography by Greg Para. Written copyright permission on file).*

Figure 7.3 Chico, a blue and gold macaw. Chico, a blue and gold macaw perched on the author's arm. *(Photography by Greg Para. Written copyright permission on file).*

help rescue parrots and bring them into the community to help others and to bring joy. So, in 2013 on January 4th, my youngest daughter's birthday, I founded Sarasota Parrot Conservatory Inc. www.sarasotaparrotconservatory.org.

How new is avian therapy?

My first thoughts on helping show others how healing working with parrots is was to look who was already doing this, and seeing if I could copy what they were doing. I found several individuals who had discovered how parrots drastically improved the moods of trauma victims. One organization, Serenity Park, in Los Angeles, California set up a small number of aviaries on the grounds of the local Veterans Administration Hospital, so veterans could volunteer to help feed, care, and interact with the rescued parrots. There was no formal process and documented case studies, so I realized this was new ground. Serenity Park is no longer operating. A story was written in the *New York Times* about its work (Siebert, 2016).

A nonprofit organization called Northwest Rescue was started by veteran Chris Diggins. He quickly discovered that other veterans found peace working with parrots. He opened another nonprofit organization dedicated to matching people with abandoned parrots and adopting them out. His organization Parrots for Patriots was founded in 2015 and is in Oregon (Parrots for patriots, 2022).

Once I founded Sarasota Parrot Conservatory in 2013, we had volunteers willing to help me bring parrots out to heal and bring smiles. We have a program for senior citizens and routinely bring parrots to senior centers to bring joy to the seniors, especially residents with Alzheimer's disease. It is documented that many Alzheimer patients somehow remember music and pets. In her article, The Magic of Pets (Allen, 2021), Kathleen Allen points out by their very nature, pets do not judge, and they are not critical. And for someone with dementia, those qualities make them a good companion. Their very presence can help reduce the effects of dementia—anxiety, agitation, irritability, depression, and loneliness.

I have found for myself and others I have worked with that parrots help reduce the effects of PTSD—anxiety, agitation, irritability, depression, suicidal ideologies, and loneliness. However, I realize that well-designed, controlled studies will be needed to study the potential benefit of parrots in reducing symptoms and disability caused by PTSD.

My original goal was to get people with PTSD to our sanctuary to feel the healing experience our parrots could offer. Because of our work in the community, we have had many news articles written, been on television many times, and are quite known locally for our work through word of mouth. We were contacted through one of our articles in the *Sarasota Herald Tribune* by a therapist with over 30 years experience with animal therapy. We were excited as she volunteered to assist pro bono. We set about determining a process to conduct intakes, track progress, and measure outcomes. Putting people and parrots together and letting the magic happen is the easiest transition in the process of avian therapy.

Establishing an avian therapy program

As a beneficiary of avian-assisted therapy of my own design, I was fascinated by the thought of helping others with PTSD. My first inclination was to go where I was comfortable—working with other military veterans. Since the inception of the program, along with military veterans, we have worked with other individuals who suffer from PTSD including first responders, domestic violence victims, and sexual assault and childhood trauma victims. However, we still have no formal procedures or guidelines for an avian-assisted therapy program.

Originally, I believed that in order to gain any credibility with a program, we would need licensed therapists, an intake process, a case management program, and measured outcomes. It took a long time to gain any traction with establishing a program and just when we got a licensed therapist—a licensed clinical social worker—as a board member, we had to relocate our rescue. Then COVID-19 hit stopping us in our tracks. This is not the end of the story.

Those with PTSD have found that our lack of a formal intake process to be less stressful than the sterile environment they have been used to for therapy and mental health care. With time, we will certainly develop a formal process that will encompass all aspects of therapy, case management, and measuring outcomes. For now, we are blessed to be able to help those with PTSD with our rescued parrots. They reported their experience as inviting, magical, and quite exhilarating. Most have admitted they have never been this close to a parrot let alone a large macaw. It is quite exciting to watch the look of awe in someone's eyes when they are introduced to a large parrot. However, many of our best parrots for therapy are small to medium size and are of several different species.

It is vital that the scientific community develop a controlled avian-assisted therapy protocol, so that it can be replicated around the globe. We continue to rescue parrots offering a forever home and bringing in members of the community to experience, firsthand, the therapeutic value of working with parrots. We have partnered in the community with several organizations who primarily deal with people who have suffered trauma. They have begun to have us bring parrots into their existing programs to add an additional layer of healing experiences.

There is a myriad of challenges to establishing a program, and a sample of issues has been provided; see Table 7.1.

Creating an avian environment

In order to either develop a program for avian-assisted therapy or adopting a parrot for your own therapy, two very helpful resources are available. The first is the Pet Advocacy Network formerly known as PIJAC (Pet Industry Joint Advisory Council). Here is their stated mission: *Our mission is to represent the experience and expertise of the responsible pet care community to legislative, regulatory, and governing bodies.*

There are many informational resources here including downloadable resource guides on caring for your parrot (Caring for your parrot, 2022). I initially took their avian care course to get more familiar with parrot care which is vital to the health of your parrot.

The other organization is the American Federation of Aviculture. They not only have a lot of resources to assist you in caring for your parrot but also have actual courses on parrot care, basic and advanced (AFA fundamentals of aviculture online courses, 2022).

I took both courses and highly recommend them. In addition, you will see with both organizations all the legislative issues regarding standards of care they are lobbying for.

Once you decide to become more familiar with parrot care, you can then determine what type of cage or flight to get your parrot, where it should be placed, and how you should interact with it including its diet. Parrots, formerly classified as psittacine live a long time, which may be considered an advantage compared to the life span of dogs, horses, and the other animals that have been used for healing purposes. Your largest parrots live an average of 70 years and the smallest around 25 years, so losing a supportive animal to longevity is not typically an issue with parrots.

Table 7.1 Challenges of establishing an avian-assisted therapy program.

Funding
Constructing proper housing for parrots
Finding qualified professionals with animal therapy experience
Providing proper nutrition to the parrots
Developing written protocols for therapy, visitors, and animals
Providing veterinary services
Volunteer program
Marketing program

Additional research and educational material regarding all aspects of avian care can be found at Dr. Susan Clubb's website (Rainforest clinic for birds & exotics, 2022). She has decades of experience as a veterinarian and runs the Rainbow Clinic for birds and exotics in Loxahatchee, Florida. She was also a major contributor to the avian care course given by the Pet Advocacy Network.

Finally, as we explore the endless possibilities animals can offer sufferers with PTSD, we must do our part to improve the state of animal welfare. We can all do our part by doing our homework before purchasing or adopting a pet, so that we can offer the best environment for both the animal and ourselves. Unfortunately, when it comes to parrots, our current legislation is still adjusting to standards of care and can lag behind on the enforcement of animal abuse situations—which is why parrot rescues are always at capacity. People do not always fully understand what they are getting into with a parrot, and thus, organizations like ours (Sarasota parrot conservatory, 2022) are always full. You can visit the American Federation of Aviculture website (American Federation of Aviculture, 2022) to see all the legislative issues they are pursuing, by state, and what you might do to help with animal welfare.

References

Adams, J. U. (2019). *Birding with benefits: How nature improves our mental mindsets.* https://www.audubon.org/magazine/winter-2019/birding-benefits-how-nature-improves-our-mental.

AFA fundamentals of aviculture online courses.(2022). https://afabirds.org/2018_WordPress/afa-fundamentals-of-aviculture-online-courses/.

Allen, K. (2021). *Alzheimer's disease: The magic of pets.* https://www.brightfocus.org/alzheimers/article/alzheimers-disease-magic-pets.

American Federation of Aviculture. (2022). http://www.afabirds.org.

Caring for your parrot.(2022). https://petadvocacy.org/wp-content/uploads/2022/01/Parrot-Care-Sheet.pdf.

Collar of Duty: Greg & the Parrots: A U.S. war veteran diagnosed with PTSD finds solace in rescue parrots Bella and Chico.(December 15, 2017). https://trakt.tv/shows/collar-of-duty/seasons/2/episodes/1.

Parrots for patriots.(2022). https://parrotsforpatriots.org/.

Pepperberg, I. M. (2008). *Alex & me: How a scientist and a parrot discovered a hidden world of animal intelligence—and formed a deep bond in the process.* Harper.

Rainforest clinic for birds & exotics.(2022). http://www.susanclubb.com.

Sarasota parrot conservatory.(2022). http://www.sarasotaparrotconservatory.org.

Siebert, C. (2016). *What does a parrot know about PTSD?* New York Times. https://www.nytimes.com/2016/01/31/magazine/what-does-a-parrot-know-about-ptsd.html.

CHAPTER 8

Animal-assisted therapy for posttraumatic stress disorder in sexual trauma survivors[1]

Simone Swartzentuber Emmons
Service Dog Strong, Portland, ME, United States

Personal story

I am Simone Swartzentruber-Emmons and I am the founder of Service Dog Strong, a Maine 501(c)(3) nonprofit organization whose mission is to provide service dogs (SDs) to sexual assault survivors and veterans with military sexual trauma who have been diagnosed with posttraumatic stress disorder (PTSD) due to sexual assault (SA) and fund the training of the Survivor and their dog. You can find out more information from our website www.sdsmaine.org/, www.simone-emmons.com/ and by reaching out to Servicedogstrong@gmail.com.

My personal story and the creation of SDs have been a lifetime in the making, not just one single event. Many events occurred that eventually propelled me to want to take a huge chance and reveal very private secrets of my life to the general public to support the goal of helping other individuals, both civilian and military, who have been in similar shoes and who may have PTSD from sexual trauma, whether incurred during military service or not, be able to get help for their PTSD the same way combat veterans who have experienced PTSD have been able to get help via an SD. It took many hours of recovery work and self-examination to get to where I am today, and I strive to help others end their constant battle with PTSD and find sustainable solutions to relieve and address PTSD.

I share my story only in the hopes it will help others. There is a great need for increased awareness of, and support for, the wonderful connection that Survivors of SA and SDs can have and the healing that can come from

[1] I will be using the terms "Survivor," "victim," "individual," "patient," and "handler" interchangeably in this chapter referencing a person diagnosed with posttraumatic stress disorder (PTSD) due to sexual assault (SA). SD will stand for service dog.

Animal Assisted Therapy Use Application by Condition
ISBN 978-0-323-98815-5
https://doi.org/10.1016/B978-0-323-98815-5.00003-3

that connection. I share my story and experience to help increase community support, and I want to change the public's view of SDs to a more accepting and understanding one. I share to raise awareness that PTSD results from many experiences other than combat and that SDs are perfectly equipped to help relieve PTSD, no matter how incurred.

I joined the Army two years after I graduated from high school. I joined during a time of war when the government was running its "Global War on Terror" after 9/11. I was continuing a family tradition, and I was proud to be going into what I thought would turn into a lifelong career for me. After working full time and side jobs, trying to afford college while making a living, I was looking forward to Army life. My plan was to learn some professional skills or earn some money for college if I decided to serve only one enlistment.

I ended up having a much shorter Army life than I had imagined. I was quickly introduced to Army life courtesy of a violent rape by a fellow soldier within a year of enlisting. I tried to stay silent about what happened, I was ashamed and the culture was very unforgiving to women who came forward and reported they had been raped. I was afraid that the rape was really my fault and I had somehow deserved it, since it didn't seem to bother my attacker. I would see him daily, and he would point and laugh and harass me to the point that it became a normal way of Army life for me. My work performance went downhill fast, I was never offered any help, and I was afraid. I had also suffered a concussion from the rape but did not receive medical attention. Instead, I was put on extra duty for coming back to the barracks late after I was raped. To receive medical help for my injuries, I would have to reveal what happened and the way the military environment was and the way soldiers would talk about persons who did reveal a rape. It was so disheartening and negative, and I felt it was best to keep my mouth shut and act like nothing ever happened even though this was impossible to do since I had suffered a severe trauma. I decided never to tell anyone about this, so that I would never be put in the situation of retelling my story and being the woman who "ruined" his career or would be called a "slut" or other derogatory names, which was all I ever heard when rape was mentioned.

Much of the military life, individuals and command seemed to have it out for women who came forward. I couldn't even process what had happened to me. I thought the military took care of their own and everyone had each other's back. Instead, I was pushed out of the military without legal representation and given an ultimatum to sign the discharge paperwork or go to "the brig." I was young and didn't know what they were doing; they labeled me with a "borderline personality disorder" diagnosis, while all the while I was actually suffering from PTSD from the rape. A doctor that met me for less than 10 minutes claimed I had this diagnosis, and it was printed eternally on my most sensitive Army paperwork and discharge. It made a huge impact. It was a slap in my face when I really needed medical attention for my physical wounds, counseling for my psychological wounds, and legal justice for the assault I had endured. Rather, I got a one-way ticket to the streets. I was kicked out to the streets, with a one-way ticket back home with no health care, no job skills, no savings or retirement, nothing but the trauma I had been trying to stuff away. Years later, I discovered articles about the overuse of the diagnosis of "personality disorders" for victims of SA who, in actuality, had PTSD. This was an amazing turn in my life that offered me hope and solidified my internal idea that I was not really the person I was labeled on all my Army paperwork that made me feel so ashamed. I was not the only one this injustice was done to! I was not alone! It didn't make it right, but it was the beginning of a pathway to healing my trauma that had gone untreated for 10 years. These are some articles and a hearing transcript from Congress discussing this "weaponized diagnosis" of SA Survivors and the military: Personality Disorder Discharges Hearing before the 112th House of Representatives (2010), Denver Post (2012), Losey (2016), Caplan (2015).

Upon discharge, I had to work many minimum wage paying jobs to afford necessities. Without medical treatment or therapy, I was fighting to survive and the trauma from the rape and PTSD symptoms had become ingrained in my life. This is common for people with PTSD if they are unable to seek therapy soon after experiencing trauma; the symptoms can become worse and destructive.

I was drinking and drugging to self-medicate to deal with the unspoken trauma and the abrupt way I was left to rebuild my life. It was not until almost 10 years later that I was able to find counseling after I married an Army veteran, I was able to take advantage of the free counseling that was offered to dependents at a local Vet Center. I didn't even know that my service would "count" for anything, and I never told anyone I was in the Army anyway to avoid unwanted memories and unwanted questions. But I had been seeking counseling for learning to live with someone who has PTSD. Before this counseling session, I had only heard of PTSD in the combat scenario. I had gone to the local Vet Center to try to help my husband with his PTSD, not knowing that I, myself, had PTSD as well.

I remember my therapist's words exactly, "I don't want to hear about your husband's military service, I want to hear about yours."

I had always tried to keep my time in the Army a secret and never shared it, even on applications. I had created a 10-year distraction of risky behavior, self-medicating and overworking to try to do anything to avoid bringing up these confusing and sad memories. My life's dream was taken from me with that one incident and the aftermath. I had never entertained the thought that it could have somehow not been my fault.

I finally told my therapist what had happened to me. It all came out, and instead of hearing that I was a horrible person and "asked for it," I was told that I was raped and I had PTSD. This was the truth that finally came to the surface to propel me to healing. Once I knew why I had been molded into the person I was that day in that chair, my life since the military all seemed to make sense. I was left with no guilt about my decisions. I saw that I was acting out of fight or flight for many, many years due to circumstances beyond my control. I learned that my PTSD was treatable, and it became my new obsession to take back my life that the military had taken, not only the year that I went through enlisted in unsafe conditions but also the 10 years after where I was living a crazy, unpredictable PTSD lifestyle. I wanted to take back all those years and change it all, but all I could change was the present and working for a better tomorrow.

I then made it my mission to find the best ways to deal with my PTSD. It was a long search that eventually led me to finding a lasting solution to PTSD symptoms through an SD. But I do feel that each therapy I tried added to the next. Healing from any trauma takes time and must accompany a strong desire to improve your life. I started with common talk therapy and went to cognitive processing therapy (CPT), cognitive behavioral therapy, and eye movement desensitization and reprocessing therapy. Yoga, Kundalini yoga, learning to meditate with Transcendental Meditation (www.tm.org), body therapies, massage, Reiki, acupuncture, hyperbaric oxygen therapy (HBOT) (oxygen chamber), and Taekwondo Martial Arts classes also helped reduce my symptoms until I found my SD. Exercise was a huge positive addition to my healing and new positive lifestyle along with eating a "clean" diet. I found my PTSD symptoms are always less when I have adequate physical activity. PTSD is literally felt in your body, so giving your body something physically positive to feel is a welcome break. Due to wanting to live an addiction-free lifestyle and as close to nature as possible, I was willing to try any noninvasive or non-pharmaceutical therapy. My continual searching for PTSD solutions led me to looking into SD therapy. It was different than any previous therapy I had tried, and thus was very appealing. It seemed to be very beneficial and promising, so I decided to make way for a dog in my life and decided to adopt an SD.

I was still feeling uncomfortable in social settings and would get easily disoriented while shopping. I was still left with unwanted PTSD symptoms and began to research SDs as a natural approach.

I found trauma begat more trauma in my life until I took active chances to make positive changes. A big influence to me living a more positive life and having a more positive outlook was being accepted into the SD program that gave me my SD, Gunnar. With that act of kindness, my heart opened some to the idea that things can go as planned. The SD program changed my life and gave me back confidence in social situations. I felt like my SD by my side was my "ticket" to anywhere I hadn't been able to go before. I was able to take back my life when I got my SD. The ability to take my personally trained SD by my side anywhere was reassuring and comforting; I never had to be alone. My SD would be able to get my attention when I dissociate or have a panic attack. It was a freeing feeling. I was able to take the lessons of owning and training an SD, use them to become stronger and not rely on my SD as much anymore. I feel so much more confident now to face many situations on my own. I have learned that I am the one in charge of my SD, and the secret is not the SD, it's me. My whole attitude has changed and I feel in charge again, like before I was raped. I learned ways to deal successfully with my PTSD symptoms with help of an SD. I was so changed by having Gunnar, I felt that everyone with PTSD needed an SD. In researching military sexual trauma (MST) more, I saw that I was certainly not the only one who suffered like I did. In researching SDs, it was hard to find organizations who helped the MST population. I felt that this was a severe injustice to MST Survivors. The anger from seeing a lack of services for MST Survivors along with the yearning to pass along this feeling of strength to others. This next step in my healing journey was so strong in me, it overcame my fears and hesitations and propelled me to start Service Dog Strong and share SDs with others because it had been so beneficial to me personally. I took a chance and it was an extremely freeing moment to share my story in hopes of gaining recognition and support.

Fast forward a few years later after sharing my dream of starting a nonprofit online, Maine has its first nonprofit organization that provides SDs to Survivors of SA and veterans who have experienced MST. I hope to see many other nonprofit organization follow in our footsteps to create better solutions to PTSD that are sorely needed across the world.

Trauma is trauma; no matter what events took place to result in the PTSD, what remains the same is that our bodies and minds react to try to protect us after that trauma. PTSD is experienced especially by those who did not get immediate help and counseling to process the trauma in a healthy way. This can be even more so for Survivors of SA because SA is so stigmatizing and there is still not enough public knowledge, support, and outreach. SA is still a taboo subject. I want to bring the subject to the forefront and talk about it. The more we talk about it, the more we create a safer environment for Survivors to come forward, and, also, for perpetrators to come forward and seek help. No one should feel ashamed to ask for help.

Benefits of service dogs for survivors of sexual assault diagnosed with PTSD

Key points for canine-assisted therapy for PTSD in sexual trauma survivors:

- Bond between an SA Survivor and an animal can become closer verses a human-to-human bond for those individuals suffering from trauma inflicted on them by another human
- Tactile stimulation by petting helps regulate Survivor body and emotions
- Owning SDs helps promote physical activity, thus helping PTSD symptoms
- SDs have no adverse side effects
- SDs are a sustainable PTSD solution
- SDs can be combined with all other PTSD treatments
- SDs can make Survivors feel more comfortable in close quarter situations by acting as a physical buffer
- Owning an SD can provide a sense of purpose

One big factor, I believe, is SDs are not human! For an SA victim, it was a human who inflicted trauma on them and, in my case, both men and women were my enemy during my trauma, thus I became untrusting of all humans. A dog is easier to bond with for this reason. A bond between human and dog can, in turn, open a door to be able to share connections with others. We can see this security build as the Survivor is able to trust again and be in public again. The trust between SDs and Survivors of SA is very strong for the reason we all know in our hearts; animals are easy to bond to. For people with trauma around humans, it just makes sense that being paired with a dog can be the start of a new life for some. Having trust again, even with an animal, can be very impactful. The goal is to develop a strong dog and handler bond. Research shows us that SDs are a viable option to offer SA Survivors that can be utilized along with other traditional therapies and not conflict. SDs have no side effects, so they are very safe as long as the Survivor has made adequate life changes to accommodate a dog and is able to care for it properly.

Service Dog Strong has concluded from our graduating class surveys that we have found an overall reduction in PTSD symptoms in participating Survivors upon graduating our class and thus owning a fully functioning SD. During our time in operation, we have not had our Survivors return surveys indicating symptoms getting worse. Some symptoms have stayed the same, but mainly have improved in the short few months we monitor them during the class. We have never conducted further surveys from years out from owning their SD. We feel that further controlled studies on SDs and Survivors are warranted. As the bond develops over the years of owning an SD, the reduction in PTSD symptoms becomes greater and greater.

I endured a violent rape when I was a soldier in the Army. I did not come to terms with nor speak of it for almost 10 years. My search for PTSD answers and help with symptoms kept me going. I knew life didn't have to be so difficult, but I did not know exactly how to reach this, so I never gave up in my search.

I found help for my PTSD in many methods and treatments. My search for recovery and relief of symptoms led me to an ongoing pursuit for self-improvement and coping skills. I utilized many therapies and each one had their benefit, but until paired with my SD, I did not notice a sizeable improvement in my symptoms. There was nothing more powerful to me than my SD. I benefited from a wide variety of therapies, each adding something to the next: EMDR or bilateral therapy; CPT; individual traditional talk therapy; Transcendental Meditation; yoga; acupuncture; HBOT; electrotherapy; chiropractic care; exercise; and eating a "clean," organic diet free from refined sugars and artificial ingredients. I was able to use these all along with my SD. Diet and exercise helped my body and mind maintain an even composure. The more exercise I got, the better I would feel. The better I ate, the less my symptoms appeared or possibly they were easier to handle when my body felt at its best. I would notice symptoms increase when consuming refined sugars or not exercising.

Having my SD encouraged me to maintain a regular exercise schedule with daily dog walks. With my daily dog walks that were "for the dog," I began to feel better.

Why do SDs work for survivors of sexual assault?

Victims of SA tend to have significant mistrust of other humans because of the way trauma was inflicted upon them, trauma inflicted by another human, and possibly using physical force. As previously mentioned, dogs can bridge the gap between mistrust and trust. They often open Survivors up to the possibility of trusting again.

Imagine that you are paralyzed with fear. You can feel it in your body strong enough to keep you from leaving your couch. You are locked in your home and too afraid to leave. This is not being able to reach a consistency with comfortable feelings nor enter a comfortable state during the day enough to feel at ease and heal your nervous system. When you are hyperalert all the time, you are unable to reach a restful or comfortable state that allows you to put in recovery work that is essential to healing. We all understand that if a person's system is in fight or flight, the person is unable to make rational choices or make good decisions, down to stopping their ability to learn. Your mind goes primal, and in that primal mind, healing and moving on from trauma are stopped. Those with PTSD can get stuck in this kind of lifestyle for years without treatment. They "learn" to live in fight or flight mode constantly. It takes a toll not only on one's mind but also on body as well. This is what makes PTSD so deadly; the continual fight or flight mode that is constant on a daily basis. This fight or flight can become engrained in a person as an habit or an unhealthy adaptation to the pain they experienced. Those who have not been able to deal with their trauma yet or a person who is unable to calm their system down enough to be open to healing work will suffer. Until the Survivor is given adequate resources, time, and support to calm their systems down, they will be unable to break the fight or flight lifestyle that becomes habit. Here is a study on learning and memory when under stress (Vogal & Schwabe, 2016).

For me, PTSD is a daily battle to maintain my cool because my autonomic (sympathetic) nervous system is always on alert. I found this out in a surprising way. I was given the opportunity to attend a three-day Kundalini yoga and meditation retreat. I had decided to attend the three-day retreat without bringing my SD and had made arrangements to leave him at home. I had made the decision to truly take some "me time" and attend without the retreat without the assistance of my SD. Without the help of an SD at the retreat, I was unable to calm myself to a point of relaxation nor felt truly comfortable. It took till the end of the three, 12-hour days of constant meditation for my body to finally feel relaxed without my SD. It took three whole days of continual meditation and yoga to finally allow my mind and body to connect and tell myself that I am safe and to be able to relax some.

PTSD can be experienced to the point where you are physically unable to calm yourself down. SDs are an external bystander if you will that can be trained to respond to the handler's every action. When I have my SD, I can reach a relaxed state much quicker than without him by my side. With my SD by my side, I feel at ease even in stressful situations. My SD can respond to my actions and intervene if necessary to help with panic attacks or dislocation. PTSD makes its victims feel uneasy and on alert 24 hours of the day. Most individuals with PTSD have difficulty sleeping, since their nervous system does not easily relax to the point of allowing the body to feel safe enough to sleep. SDs can be trained to wake handlers when they are having nightmares, or some dogs can just naturally do this when bonded with the handler. One of the reasons PTSD can be so difficult to treat is because it is not like an injury or illness that comes and goes, physiological responses are always affecting the Survivor. Herein lies the beauty of SDs to treat mental issues; the dog is with the person, just like the illness, 24/7. Unlike a therapy group or counseling session that lasts a few hours a week, the SD is not an hour or a weekly commitment, it is a 24/7 commitment; the SD is always by the Survivor's side, day and night, continually reiterating attention to the person's condition and providing reassurance and physical or emotional assistance to that Survivor. If the Survivor can get to a comfortable place with the dog, then we see the Survivor's physiological sympathetic nervous system is not on high alert and the person lives a more normal life. Dogs have the ability to lower a person's heart rate and restore the person to a quieter place, so the sympathetic nervous system can take a break. Maruyama et al. (2012) is a study showing the relationship between tactile simulation dopamine. Drescher et al. (1980) examined heart rate in response to touch.

When the sympathetic nervous system is allowed a break, the person returns to baseline and oftentimes without the need for added medication. The person may be able to self-regulate within moments of being with the dog, to a place that otherwise may have taken many hours of self-teaching on behalf of the Survivor. Animals can have this enormous effect on us humans, especially those of us who have experienced trauma. We have seen this happen with many Survivors; the ability to relax in new, stressful situations, such as the first day of our classes. The Survivors are in a totally new location to them, but they are there with their dogs by their side. Often these Survivors were homebound due to fear and isolation directly related to their PTSD. The animals have an extraordinary effect on our sympathetic nervous system, and they just do it naturally. We have seen that a connection to the Survivor's dog makes them more comfortable in new situations. The strong bond of reliance and trust between the Survivor and their dog (since the SD is with them 24/7) is a key gateway to open the Survivor up to leading a new life of making positive changes, since they are able to relax enough to allow this to happen!

The strong connection of assurance in animals creates daily positive pathways by multiple methods; the dog will be a physical block between the Survivor and a stranger, tactile stimulation, emotional bonds, all creating a physical sense of safety for the handler. The physical blocking technique relates to emotional well-being in people who have been hurt physically or have had their physical boundaries crossed. The real physical distance the dog creates is often enough for many Survivors to regain the confidence they need to put themselves in public situations again, when they had previously withdrawn from them.

Repeating daily tasks, such as daily walks and daily training is also beneficial to Survivors with PTSD due to the physical aspect (physical activity is proven to help depression symptoms). Also, by learning new tasks, it challenges the mind just like a crossword puzzle, but with more concentration since you are working with a changing medium—the animal. The daily act of grooming and brushing the dog teaches caring and can be a peaceful activity for the Survivor to engage in. When they create a routine, it can be very beneficial to the Survivor to have a sense of purpose and structure. PTSD can often be a debilitating lifelong disability. Creating routine and a sense of reliance (the dog needs the Survivor), in turn, creates a positive example for the Survivor themselves to create positive daily routines for themselves. The primary goal is for the Survivor to become more self-reliant and brought out of depression as much as possible. There are some cases where we see the Survivor become lifelong-dependent on having an SD. We had an applicant who had previously had two SDs. When one died or was too old for service, the Survivor felt lost and unable to stay on schedule themselves; thus, their anxiety in crowds returned and they fell into a deep depression until they reached out to our organization looking for assistance to get another SD and return to the engrained reliance on the SD. Often, when this happens, the SD performs dual roles for their Survivor. They need the dog for the mental PTSD, but also they will have a physical problem that the SD is trained to assist such as a balance problem or seizure alert dog.

In cases where a Survivor has physical limitations in addition to PTSD, it is understandable that they would need an SD to assist for a lifetime once they become comfortable with their SD and it has shown improvement in their daily lives. For other Survivors, depending on multiple factors (since every person is an individual and you cannot group all Survivors together), they may only need to rely on their SD for a few years when great improvements are made. This is almost always in combination with other therapies, such as counseling, body work, alternative therapies such as chiropractic care or acupuncture that bring the individual in closer connection to one's body. Bridging the gap between what happened to the Survivor's body and who they are now, and reinforcing their body is not in physical harm at this moment can bring much success in treatment and healing. SDs can also bring that physical sense of calmness to the Survivor by daily physical interactions with their SD.

Physical tactile stimulation creates calmness and a quieter sympathetic nervous system. It can help with anxiety attacks and can be very helpful for PTSD. These articles—(Lloyd et al., 2019), (Gravrok et al., 2020)—support such observations and the idea that SDs so then are very beneficial to persons with PTSD.

How to approach a survivor of sexual assault and service dog team in a medical setting?

My approach to individuals with PTSD is based on Eastern philosophies where we are made up of mind, body, and spirit. This can be considered trauma-informed care or trauma-informed approach. I believe you cannot treat PTSD (or any mental health challenge for that matter) effectively psychologically without treating and addressing the person's needs as a whole. Treatment is more effective when the individual has care for their body as well as their mind and spirit. My definition of "spirit" solely means the person's connection to a higher power. This can be seen in many ways, such as the traditional connection to a God or nature or can be seen as having a purpose in life. Especially for people with PTSD, having a connection to other peers, or a higher power, or just having a mission in life to keep their spirit up is very important. Dogs also have a soul or some kind of connective energy that is a key distinction to SDs verses traditional therapies. Dogs have a bond and spirit energy to share. You feel this upon taking over the leash. When a Survivor is paired with an SD, they both create a connection and bond. This bond can be a tremendous reassurance to a Survivor who may not have ever felt comfortable connecting to another human posttrauma. As doctors, you are mainly treating the person's physical symptoms, but I think it is becoming apparent, even in Western cultures that any medical treatments are more effective when an individual is treated holistically, noninvasively, and if the patient's whole life and mental situations are taken into account and respected. Being treated for ailments noninvasively can be very important to SA Survivors, since there has already been a history of trauma that has occurred to their physical body. SDs are an absolutely noninvasive treatment. SDs hold a special bond between themselves and their handler that is like a connection to another human but oftentimes stronger.

Dogs are unconditional lovers. Any human, even the most respectful ones, can sometimes hurt us or portray mixed signals from time to time. SDs avoid all that unwanted body language or potential for unwanted abuse, which can be shown by us humans; the dogs directly connect to the handler by a strong bond that is felt deeply, since a Survivor can trust the SD which will not violate their personal space in a negative way. The trust bond between a Survivor and SD can be felt even if the two have been working with each other for some time. By addressing the person's mental reactions, body reactions, and respecting the spirit of the person, or purpose, this increases their will to live and recover. This can also be very closely connected to trauma-informed care. Trauma-informed care includes addressing the patient holistically is key, asking if they feel connected to a higher power or if they have peer connections or a sense of purpose in life. This aspect can sometimes be missed in interactions with care providers and is very important to patients with PTSD and anyone who has experienced any form of trauma. Since it is impossible to know who has experienced trauma in their life, we must approach each patient with an empathetic trauma-informed approach.

How do you approach patients who have experienced SA using trauma-informed care? Who is a victim of SA? Since around 5% of SAs are reported, I argue that care providers should treat every patient as a potential Survivor and approach them all with care and awareness. Some victims will not disclose that they have been assaulted, so the best practice is to use trauma-informed care models for every patient.

I don't believe trauma-informed care should be found only at one specialized office, but every care provider should strive to provide it to every patient they see.

Knock before entering the room. Ask before touching. Explain all procedures thoroughly and allow for the patient to be given choices in their care, so that they feel and can be more in charge of their own body. These practices help give the patient control over their own body when they are allowed to make their own choices about what happens to them. Taking away choices in care can feel like an attack to a Survivor of SA. Not given choices or not having procedures explained properly can cause anxiety and make the patient feel like they are being forced into an outcome or decision about their own body they were unable to have a say in. This is especially true for SA Survivors; they can relive trauma unconsciously or consciously when they are touched without being asked, or from being told what is going to be done to their body without being asked their consent or given options. The switch from simply telling someone what is going to happen to them as compared to giving patients options and letting them decide what happens to their body is ultimately giving the patient back control. A lot of medical scenarios put the patient in a box where they have no alternative options of what happens to them. But giving the ability to let patients have as much say and choice as possible over their own body and allow adequate time for questions are most important to prevent further trauma.

Trauma-informed care is also appropriate SD etiquette. I have actually been asked by a physician of mine "Is there anything special I need to know about your SD?" Although this physician might have had the best intentions, people with SDs want to be treated as if they don't have an SD with them. Having the care provider's attention solely on the individual is important. It may be challenging to not treat the dog like a pet, however cute they may be, but please, out of respect for the handler, leave the "pet" talk behind. The handler has their dog out of necessity and because of their disability. The handler would much rather not have to bring their dog everywhere or have the disability that is supported by the SD in the first place. Psychiatric SDs, for example, call attention to individuals' unseen disabilities. Most individuals do not want others to know they have a disability. The SD is a "sore thumb" when it comes to the concealment of "medical equipment."

Female-identifying Survivors of SA with SDs may have a stereotype to deal with. I have been asked if my SD was actually mine when a store manager looked at me and seemed to assume I did not have a disability since "I have all my limbs." I have also been approached by complete strangers and have been told to "Thank my husband or boyfriend for his service. He must have been in the service. This must be his SD, right?" The general public's view on SDs seems to indicate a lack of knowledge of who SDs can serve. Unless you have a visible physical disability or are a large man who looks like a combat veteran, the general public, even some physicians, seem to have this preconceived notion that SDs are only for the visually impaired, amputees, or combat veterans. But this notion is false because we know firsthand, having seen the transformations happen, that psychiatric SDs help those of us who have invisible psychiatric disabilities like PTSD.

The beauty of the SD as medical equipment is that it can be combined with other medical treatments, pharmaceutical, counseling, therapy, or other holistic treatment and it will not be an interference. SDs can be considered an additional therapy to ongoing treatments or be used solely with a care team and be the person's primary method of treatment. I love the use of SDs to help manage PTSD symptoms because it is a daily form of treatment. The handler uses their dog daily and has daily interaction with the dog, thus bringing attention to the individual's need for the dog and attention to their disability. Also, it brings a daily "preventative medicine" to the Survivor's life. The treatment arrangement options are as individual as the individuals. I also love the fact that SDs can be used in conjunction with other therapies and modalities. Having the dog by their side each day is a daily form of treatment if the individual chooses to use the dog properly, continues working their training program, not letting the dog turn into a "pet" but adhering to the strict SD requirements. To do this, the handler must consistently train and utilize their SD in public, so the dog does not forget their drills and commands. An SD may be utilized by a patient who is unable to utilize traditional pharmaceuticals due to health conditions or even a personal choice not to medicate (while pregnant, for example).

Connection is an important key. Having a connection to another being, whether human or animal is important for strong and solid mental health. For a person with PTSD, having a strong connection to a friend, counselor, or animal can, in itself, create a safe place for the person to come out of fight or flight mode (a hyperactive sympathetic nervous system reaction) and allow them to relax enough to create new, positive brain connections. Animals have a very strong connection to us; we often treat household pets as "one of the family" despite being a different species. Often a person with any sort of trauma can feel this special bond more quickly and is more open to sharing a bond with an animal before a human. As we know, animals are nonjudgmental creatures, and this can create a safe place for a Survivor to trust again and make connections with others. We hope this will help the person feel open to making connections with others again in the community, family, and/or peers. With regard to Survivors of SA with PTSD, due to the hyperactive nervous system, each day is faced feeling like each action you take is a life-or-death action. Survivors of SA tend to have significant mistrust toward other humans because of the way trauma was inflicted upon them, trauma that was inflicted by another human, often using physical force. Dogs can bridge the gap between mistrust and trust. They often open Survivors up to the possibility of trusting again. Dogs can fill that mind-body-spirit or emotional gap that can exist in sole therapies, like only using pharmaceuticals or one therapy modality.

I would like to add an amazing publication resource for practitioners (Simkin, 2004) that is the first of its kind to document such helpful research that will increase the efficiency of care for Survivors. Taken directly from the author's website, "The only book of its kind, When Survivors Give Birth provides survivors and their maternity caregivers with extensive information on the prevalence and short- and long-term effects of childhood sexual abuse."

Service Dog Strong organization

Due to the great need for organizations like ours across the country and around the world, we continue to have many people reach out to us from around the country looking for an SD. Our mission, at this time, is to provide dogs and fund training for adult Maine residents who have been diagnosed with PTSD due to SA, resulting in trained Survivor/SD teams to meet the world head on. Applications for our program can be downloaded from our website www.sdsmaine.org/, but please keep in mind our reach only extends to the state in which we currently operate. We are only able to help Maine residents at this time. We would love to see this help be available to all Survivors who need it, without borders. We believe this will happen once more awareness is made over how important SDs can be for SA Survivors. Service Dog Strong has been made up of volunteers from the beginning; we have never had any paid employees. We could certainly have used someone devoted to managing the program and others devoted to fundraising in a full-time capacity, but our priority has been raising enough funds to support training classes for our Survivors and their SDs in training. It still seems that the subject of SA is too sensitive, a subject to discuss or support even when the public is continuously subjected to sexually aggressive images in the media and the representation of sexual violence by news organizations (e.g., see (Schwark, 2017)).

Some are still uncomfortable with supporting or having an uncomfortable conversation about supporting a nonprofit organization that focuses on those who have experienced sexual violence. Here is where we can make a difference and talk about the uncomfortable. The more we talk about sexual violence, the less taboo it becomes, and the less stigma will exist for Survivors of sexual violence.

The SD team training class is 20 weeks in-person, and students are expected to complete over 250 hours of individual training on their own in addition to the in-person classes. Just as the dogs are thoroughly screened, our applicants are as well. We must ensure that each applicant can take care of a dog, mentally, physically, and financially. Thorough screening of both potential human and canine participants is very important because the person must be at a place in their treatment and healing where they are able to take on responsibility and the canine must possess the potential for the ability to provide the sort of support that SDs are expected to once they are trained.

We at Service Dog Strong require all our participants to be currently and actively working with a licensed clinical mental health care provider or health care equivalent. During SD team training, memories may surface, and new situations may occur that can set the Survivor back if they are not willing or able to face their trauma head on and strive to work past uncomfortable feelings and activities they may experience while completing training. The training is very challenging for most Survivors because it puts them in positions and experiences they may have not been in some time, like going out to a restaurant. The SDs must be trained for a wide variety of situations, so that they are prepared to assist the Survivor in any situation. Situations can be uncomfortable but typically lead to confidence when achieved. Since our format is self-training, our Survivors must be willing to step outside their comfort zone. I prefer the hands-on approach and believe it is definitely the best method for teaching rather than having a trainer train the dog individually and then hand the dog over to the Survivor.

Unfortunately, the statistics show that one in five women experience a rape in their lifetime and one in 14 males (Sexual Assault Statistics, 2018; Smith et al., 2018).

With rates like this, SA is too prevalent. SA is a global issue down to a serious local problem. We are all affected in some way. It doesn't have to be this way. We can break the silence, encourage awareness, and choose to not be uncomfortable but be open to helping and speaking up. We all have had some form of indirect or direct contact with *sexual assault* or know someone who has been assaulted, so let's feel free to speak about this subject in a more open way to find solutions and to ask others if they are needing help. SA can be something we are not ashamed to talk about and that there can be help readily available for. The answer to helping this large population of Survivors can be sustainable SDs!

Not only are SDs an environmentally friendly solution but also can be even more sustainable when the dogs are pulled from shelters and complete the training. Service Dog Strong tries to rely primarily on shelter dogs for our classes, but this can pose potential issues from not knowing the dog's possible past trauma of their own. We also rely on a few local, humane, and ethical breeders when we are not able to find the right dog for a Survivor through a local shelter in a timely manner. But this is done in a last resort. Each dog, whether from a shelter or from a breeder, is thoroughly evaluated by our trainers to ensure proper temperament to become an SD. Once they are successfully screened for temperament and nonreactivity and approved, they are paired with the Survivor. The Survivor and dog then take some time to bond and then start class.

We operate differently than many other SD nonprofit organizations in that we don't just deliver a fully trained SD to an individual, we fund the Survivor who enters a class where they learn to train the dog themselves alongside an instructor. Our training model involves the Survivor adopting (at no cost to them) a shelter dog that we hand select for their individual needs and pair with them, and the two enrolling in an SD training class led by a highly trained instructor. Our class transforms the dog from a shelter dog with no skills, but plenty of potential, to highly trained medical equipment for the Survivor after completion of the group training class and hundreds of hours of the team training together at home. The Survivors and their dogs attend class once a week, they learn and practice skills in class, then train with their dogs at home logging their progress and hours of training in a notebook.

For some disabilities, visual impairment, for example, the handler is unable to train the SD themselves. However, for psychiatric SDs, involving the Survivor in the direct training of their SD is very beneficial and preferable. Teaching the handler to learn how to train their own dog themselves creates personal involvement in their own recovery and treatment plan. Being involved in their own treatment can create a sense of worth and a sense of purpose. It can also carry some additional pressure that not every individual with PTSD is prepared for. The training can prove to be very intense at times when training for the person's individual triggers. They will have to be prepared for uncomfortable situations until the dog is fully trained and able to assist them fully. During the training, the Survivor will have to have solid mental health support. Again, we require Survivors to actively participate in individual therapy while attending our class and we encourage them to continue afterward as needed.

Survivors must be able to plan goals and schedule each day on their own time to train individually with their dog. A secondary goal of the program is teaching Survivors to be more self-reliant. In some cases, upon receiving an SD or shelter dog in training, the individual is propelled to practice self-care alongside caring for the dog. Feeding their dog every day can remind Survivors to feed themselves. It can encourage self-care this way. PTSD can inhibit people from getting out of bed sometimes due to fear and anxiety, but with the addition of a dog, the mission of getting up and taking care of them, such as taking them for a daily walk can create a new sense of motivation, in turn helping the Survivor create a new motivation to live and keep going. We have witnessed Survivors return to work after receiving their SD, since they are now able to bring their dog to work with them. We commonly see Survivors gain confidence with their dog and are more able to return to public places or go to appointments once again feeling confident in closed spaces where the individuals are alone, such as a typical doctor's exam room where the door is commonly shut. The addition of an SD to a Survivor's life can mean big impacts in small ways each day. The comfort of having the dog to rely on in close quarters situations can be very reassuring to the individual's sense of well-being.

Involving the Survivor in training is more cost-effective than the typical SD training. Since the handler does much of the training at home, there is no cost to the organization to train the dog and the organization is not responsible to pay an individual trainer to train each dog the required 250 hours, since it is completed at home personally and the trainer is paid for the teaching the one training day a week for 20 weeks.

The handler in this training model is completely involved in the outcome of the class. Thus, results are based on effort. This is a perfect model for real-world situations. Teaching self-reliance and hard work brings a huge sense of accomplishment at the completion of class. This is also the reason we thoroughly screen our applicants to ensure they have the determination and drive to complete the often-challenging class. We want to ensure they are stable enough to handle the pressure of daily tasks and training. Our goal and hope is that each Survivor rises to the challenge but we have also seen a few Survivors retract their applications after careful consideration of what is required. We encourage them to reapply when they believe they will be more prepared for success, and we hope that they return.

Potential drawbacks to having a service dog

Are there drawbacks or is there a "downside" to having an SD? The short answer is yes, there can be. The "downside" to having an SD is that individual results may vary. We have heard that term used before, and in the case of SDs, it is very true. Each person puts in different amounts of effort and connections to their dogs can be different, and the dogs' abilities are all different. Each dog has come from different backgrounds and genes. This affects their personality, especially if they are shelter dogs that have experienced trauma in their own life before being adopted. Animals are always subject to a certain degree of unpredictability no matter how well-trained. Each team responds differently to the addition of an SD to their treatment plan. We have seen SDs totally change Survivors' lives for the better. We have had the unfortunate events happen where a Survivor doesn't bond with their dog and does not have life-changing results. For some, the addition of an SD to their life results in increased stress. A common challenge for individuals with invisible disabilities is the extra attention they encounter when bringing a dog into public. Where they once were just another person in the crowd, now the SD makes them stand out like a sore thumb in a crowd or store. Sometimes this added attention can cause more stress, since it calls direct attention to the person's disability.

There is no guarantee that a Survivor will form a bond with their dog and that bond will translate that connection into trust for others; there is no guarantee that PTSD symptoms will subside, but when the Survivor is highly involved in their training, they have a much higher chance to succeed. It can be challenging to find the proper fit of dog/human personalities, but we see most success when the Survivor is highly involved in their training, in individual counseling, and utilizing other tools for healing.

My SD, Gunnar, has been a great tool in my PTSD-kicking toolbox that has allowed me to see my strength, step outside, and my confidence has grown.

We see most problems with success arise when Survivors choose not to train their dog properly or do not stick to consistent training. Having an SD requires daily training and tasks. This can be challenging for some, initially. The Survivor must be able to commit to lots of time and hold a strong commitment to heal. The effort will pay off when the team work together in the public, testing their skills in a real-life setting.

Conclusion

This is what help looks like from a PTSD SD that has been specifically trained for an individual by the individual themselves: The Survivor wakes up, and the dog gets up with them. The Survivor gets into the shower, and the dog sits patiently outside the curtain, waiting and listening. The Survivor gets ready for their day, the dog sitting by their side waiting and watching for any disturbance or change in the Survivor's demeanor. The Survivor is not alone anymore. The dog monitors the Survivor's behavior constantly during the day and is trained to react when disturbances occur, faster than a call to their therapist, faster than a trip to the medicine cabinet, quicker than the individual can themselves realize what is happening sometimes. And that's why PTSD SDs are a constant enforcer of the person's therapy, being as a reminder of calm ways to deal with PTSD symptoms. When SDs are used alongside counseling, the positive results are experienced quicker in those who actively participate in their individual therapy sessions.

PTSD, no matter how it was introduced to a person's life, is experienced in many ways through the body, thoughts, and emotions. Physical disturbances such as hypervigilance, anxiety attacks, flashbacks where blood pressure heightens, sleep issues, higher risk for diseases, or fertility issues because of the overworked fight or flight system are not uncommon. Insomnia, memory loss, fear of public spaces, disability, or inability to work or to keep a job, trouble with relationships, criminal activity, or dangerous behavior are additional symptoms shared by PTSD Survivors, no matter the trauma. This includes combat veterans, victims of SA, those who have been a witness to a violent crime, soldiers returning home from war, or Survivors returning home from a forensic exam after a SA. PTSD affects our bodies and minds very similarly, but individuals with PTSD can also experience additional fears related to the specific trauma they experienced. For example, most people with diagnosed PTSD have difficulty sleeping, and any PTSD SD can be trained to help with night terrors, for example, but there can be differences in the details of their trauma. For example, a woman who has been sexually assaulted by another woman might be extra fearful of other women. This is not a symptom of PTSD, but a specific issue that person with PTSD will have to address, but it does not affect an SD training. A dog can be personally trained to help an individual with their trauma-specific triggers as well as in a general way to assist in the most common PTSD issues that all the majority of Survivors with PTSD share, regardless of the trauma that caused it.

I would like to add the article by Girgenti et al. (2020) as a reference that PTSD may be found at a higher rate in women than in men.

This is new research that can help us shift our view of what PTSD victims look like and bring a greater encompassing view of whom experiences PTSD and who may need support. There is a great population who are not combat veterans who have been diagnosed with PTSD. In the current climate where women are disclosing/reporting rape or assault more often than men, we hope that soon all Survivors of all genders will feel comfortable disclosing their trauma and asking for help. The stereotypical image that PTSD is only found in veteran men does not fully convey the percentage of the population that experiences PTSD.

There should be future controlled studies of the efficacy of SDs and SA Survivors. More controlled studies are warranted comparing combat PTSD and SA PTSD with SDS.

Research shows that SDs are helpful to combat veterans with PTSD as well as Survivors of SA with PTSD. SDs help treat PTSD symptoms regardless of who the handler is or how they experienced trauma and acquired PTSD.

References

Caplan, P. J. (2015). *Military changes course on psychiatric discharge for sexual assault survivor.* Women's Media Center. https://womensmediacenter.com/news-features/military-changes-course-on-psychiatric-discharge-for-sexual-assault-survivo. Original work published 2015.

Denver Post. (2012). U.S. military illegally discharging veterans with personality disorder, report says. The Denver Post www.denverpost.com/2012/03/22/u-s-military-illegally-discharging-veterans-with-personality-disorder-report-says/.

Drescher, V. M., Gantt, W. H., & Whitehead, W. E. (1980). Heart rate response to touch. *Psychosomatic Medicine, 42*(6), 559–565.

Girgenti, M. J., Wang, J., Ji, D., & Cruz, D. A. (2020). Transcriptomic organization of the human brain in post-traumatic stress disorder. *Nature Neuroscience, 24*, 24–33.

Gravrok, J., Howell, T., Bendrups, D., & Bennett, P. (2020). Thriving through relationships: Assistance dogs' and companion dogs' perceived ability to contribute to thriving in individuals with and without a disability. *Disability and Rehabilitation: Assistive Technology, 15*(1), 45–53. https://doi.org/10.1080/17483107.2018.1513574

Lloyd, J., Johnston, L., & Lewis, J. (2019). Psychiatric assistance dog use for people living with mental health disorders. *Frontiers in Veterinary Science, 6*, 166. https://doi.org/10.3389/fvets.2019.00166

Losey, S. (2016). Military must do right by wrongly-discharged sexual assault victims, advocates say. *Air Force Times*. https://www.airforcetimes.com/news/your-air-force/2016/05/19/military-must-do-right-by-wrongly-discharged-sexual-assault-victims-advocates-say/. Original work published 2016.

Maruyama, K., Shimoju, R., Ohkubo, M., Maruyama, H., & Kurosawa, M. (2012). Tactile skin stimulation increases dopamine release in the nucleus accumbens in rats. *The Journal of Physiological Sciences: JPS, 62*(3), 259–266. https://doi.org/10.1007/s12576-012-0205-z

Personality Disorder Discharges Hearing before the 112th House of Representatives. (2010). *U.S. house of representatives hearings*. Washington, DC, USA: U.S. Government Printing Office. www.govinfo.gov/content/pkg/CHRG-111hhrg61755/html/CHRG-111hhrg61755.htm.

Schwark, S. (2017). Visual representations of sexual violence in online news outlets. *Frontiers in Psychology, 8*, 774. https://doi.org/10.3389/fpsyg.2017.00774

Sexual Assault Statistics. (2018). www.nsvrc.org/statistics/statistics-depth.

Simkin, P. (2004). *When survivors give Birth: Understanding and healing the effects of early sexual abuse on the childbearing woman*. Classic Day Publishing. http://www.pennysimkin.com. Original work published 2004.

Smith, S. G., Zhang, X., Basile, K. C., Merrick, M. T., Wang, J., Kresnow, M., & Chen, J. (2018). *National intimate partner and sexual violence survey: 2015 data brief—updated release*. National Center for Injury Prevention and Control Centers for Disease Control and Prevention. www.nsvrc.org/sites/default/files/2021-04/2015data-brief508.pdf.

Vogal, S., & Schwabe, L. (2016). Learning and memory under stress: Implications for the classroom. *Npj Science of Leaning*. https://www.nature.com/articles/npjscilearn201611. Original work published 2016.

CHAPTER 9

Animal-assisted therapy for pediatric patients

Beth L. Macauley
Department of Communication Sciences and Disorders, Grand Valley State University, Grand Rapids, MI, United States

Children and animals, what an incredible partnership. Thinking about children and animals brings to mind images of puppies sleeping beside their human siblings, children throwing a stick for their dog, feeding barn animals before school, riding a horse bareback through the woods, and even holding bunnies and guinea pigs in their lap while doing homework. There is nothing better for children than having a relationship with and responsibility for animals. Animals have been reported to improve the emotional and functional status of children since the time of the ancient Greeks (Reide, 1987). More recently, animals were brought into the health care environment to visit patients. These visitations were found to improve rapport, facilitate communication, and increase patient responsiveness and social interaction with health care professionals, patients, and other family members (Uglow, 2019). These observations led rehabilitation professionals to incorporate animals into their treatment sessions through animal-assisted therapy (AAT). AAT has been defined as a goal-directed intervention that utilizes the human–animal bond as an integral part of treatment (Granger et al., 1998). This chapter focuses on the use of AAT within the pediatric population, specifically, an overview of AAT with dogs, horses, and other animals, under the direction of licensed therapists working with infants, children, and adolescents with psychological, physical, mental, social, and communication disorders.

It is important to note that during AAT, a bond, a relationship, is created between the client and the therapy animal. This bond is very important in the treatment process, even when the treatment goals do not include building relationship or improving communication. The relationship between the animal and the client is the underlying thread across all sessions in which AAT is used as a treatment strategy.

Animal Assisted Therapy Use Application by Condition
ISBN 978-0-323-98815-5
https://doi.org/10.1016/B978-0-323-98815-5.00009-4

Dogs

Certified therapy dogs are the main animal incorporated into AAT sessions with children. Dogs have participated in AAT sessions with hospitalized children, children with mental health disorders, autism spectrum disorders (ASDs), and cognitive impairments, as well as with children undergoing physical therapy, occupational therapy, and speech-language pathology clinical services.

Hospitalized children

Several studies targeted the effects of AAT on the psychological and emotional welfare of hospitalized children. Specifically, participation in an AAT program resulted in improved mood, improved emotional state, and decreased anxiety and pain in hospitalized children, including those in intensive care units (Acerboni et al., 2013; Ávila-Álvarez et al., 2020; Hinic et al., 2019; Ichitani & Cunha, 2016; Jalongo et al., 2004; Nilsson et al., 2019).

Silva and Osorio (2000) studied 24 children with cancer who were undergoing outpatient oncological treatment and their caregivers. The children participated in three 30-min sessions of a structured AAT program. Psychological, physiological, and quality of life measures to evaluate pain, mood, anxiety, stress, depression, heart rate, blood pressure, and quality of life were given before and after the AAT program to the children and their caregivers. Results indicated that the children reported decreased pain, irritation, stress, and depression following the AAT program, while their caregivers reported improvement in anxiety, mental confusion, and tension (Silva & Osorio, 2000). Calcaterra et al. (2015) reported that hospitalized children who received AAT after surgery experience less anxiety and less pain (Calcaterra et al., 2015). A study by Nahm et al. (2012) documented that 93% of staff and patients approved having therapy dogs in the emergency room to decrease the stress and anxiety of patients, including children (Nahm et al., 2012). More recently, a study by Jennings et al. (2021) reported an increase in mood and activity levels as well as a decrease in the stress hormone cortisol in 80 children aged 2—19 hospitalized in inpatient pediatric acute care units following sessions with therapy dogs (Jennings et al., 2021).

Two systematic reviews of the effectiveness of AAT on hospitalized children studied the results of four randomized control trials, four quasiexperimental design studies, and 3 controlled clinical trials (CCT). The conclusions after reviewing these 11 studies indicated that hospitalized children and teenagers had less pain, lower systolic blood pressure, and higher diastolic blood pressure as compared to controls. There was no difference between the two groups on anxiety, stress, and heart rate (Feng et al., 2021; Zhang et al., 2021). This latter finding is intriguing because many studies have reported improvements in these areas for hospitalized children following AAT sessions. It is possible that those studies did not meet the criteria for inclusion in the systematic reviews as there is definitely more than 11 peer-reviewed research articles on this topic.

Many hospitalized children with serious illnesses and degenerative diseases may eventually receive palliative care. A review of the literature on the use of AAT with children with life-threatening and life-limiting conditions was conducted by Gilmer et al. in 2016. Conclusions indicated that AAT helped to decrease the suffering of children receiving palliative care. Gilmer and colleagues encouraged more palliative care programs to implement AAT with this vulnerable population (Gilmer et al., 2016).

Mental health disorders

For the purposes of this chapter, the area of "mental health disorders" will include a mixed bag of psychology-related services including AAT with incarcerated youth, at-risk youth, and children receiving psychological counseling or psychiatric treatment for sexual abuse, posttraumatic stress disorder (PTSD), and substance abuse. Studies of AAT with incarcerated youth target the relationship aspect of human—animal bond in the hopes that this unconditional relationship will encourage the youth to grow in their self-esteem, confidence, and decrease their problem behaviors. In fact, studies have documented that incarcerated youth who participate in AAT with dogs show improvement in social-cognitive skills such as making socially appropriate decisions. The youths even reported that they were beginning to internalize their problems rather than take their problems out on others which resulted in a decrease in altercations between the youths as well as between the youths and their correctional officers. In addition, the incarcerated youth appeared to have an increase in empathy with a more positive outlook in life (Seivert et al., 2018; Syzmanski et al., 2018). A systematic review of 20 studies that met the inclusion criteria reported that the average length of AAT sessions was 60—120 minutes with sessions occurring 1—3 days a week. Results indicated that improved mental health, emotional control, empathy, and academic skills could all be improved after AAT intervention with dogs. These improvements were statistically significant in 13 of the 20 studies. The authors concluded that AAT with dogs may improve the anxiety, stress, recidivism, and other social variables in both young men and women (Villafaina-Domínguez et al., 2020).

Studies on the impact of AAT with children who experienced physical and sexual abuse, children with PTSDs, and children with mental disorders have shown that participation in an AAT program, integrated or adjacent to their counseling, helped the children to reduce the internalization of the experiences and be more open to therapy. Many children also showed a decrease in their PTSD behaviors and were able to participate and engage more readily in counseling. These are significant clinical results in that many children who experience abuse and trauma are unwilling to be vulnerable and openly talk about what happened, their reaction, or how to understand and manage their feelings and their life after the trauma. If integrating a certified therapy dog into their treatment facilitates open communication and improvement in their symptoms, great! (Allen et al., 2021; Dietz et al., 2012; Kelly & Cozzolino, 2015; Maria Cristina Stefanini et al., 2016; Muela et al., 2021; Signal et al., 2017; Stefanini et al., 2015; Trujillo et al., 2020). An interesting study by Prothmann et al. (2006) used a pretest—posttest design to study the effects of an AAT program for children and adolescents who were receiving inpatient psychiatric treatment. A formal measure of general state of mind was given before and after AAT for one group of children and at the same times for a group of children not receiving AAT. A comparison of the two groups found that the group of children who participated in AAT had significantly higher scores across all four variables of vitality, emotional balance, social interaction, and alertness while there was no change in scores for the children who did not receive AAT. The authors state that incorporating a therapy dog could catalyze psychotherapeutic work with children and adolescents (Prothmann et al., 2006).

A systematic review of AAT in psychotherapy for children aged 10—19 years completed in 2019 investigated the findings of seven studies that met the inclusion criteria. These studies documented that AAT for this population had a positive impact on primary diagnoses and symptomatology, decreased symptoms of PTSD, anxiety, and anger. AAT was also associated with positive impacts on secondary factors including increased engagement and socialization behaviors and decreased disruptive behaviors during treatment sessions. Global functioning of the children also improved. However, there was insufficient evidence that AAT improved factors associated with self-esteem, subjective well-being, or coping (Jones et al., 2019).

Related studies documented the benefits of AAT for children who experience anticipatory anxiety and fear before dental treatment and before magnetic resonance imaging (MRI) studies. Children with significant anxiety and fear at the dentist had a reduction in symptoms when a therapy dog was present and were able to manage the dental treatment (Vincent et al., 2020). Similarly, children with significant anxiety prior to MRI demonstrated a statistically significant improvement in anxiety levels following AAT. Of the 21 children who participated in the study, 19 were able to successfully complete the MRI study. The authors concluded that the AAT resulted in a beneficial effect on patients' emotional status, easing anxiety in preparation for scheduled scans, without impacting MRI quality or duration (Perez et al., 2019).

Adolescents who received AAT during residential care following traumatic experiences demonstrated more secure attachment and less impact of parental interference than the adolescents who did not participate in AAT. These improvements allowed the adolescents to deal with the trauma and respond better to their care (Balluerka et al., 2014). Studies have also been completed on incorporating AAT into family preservation services. Families referred by Child Protective Services were randomized into a group that received AAT in addition to services and families who only received services with no AAT. Family functioning outcomes were measured using the North Carolina Family Assessment Scale for Reunification. Results indicated that all four targeted family functioning outcomes were significantly increased for the families who received ATT as an adjunct to intensive family preservation services ($n = 14$) with medium to large effect sizes, while there were no significant differences in the measures for the families who did not receive the adjunct AAT. The authors suggest that adding AAT to intervention for at-risk families can improve evidence-based clinical interventions and improve the caregiving contexts of these children (Flynn et al., 2019).

While the above studies, as well as many others, document the benefits of AAT within psychology and psychiatric services of children and youth, one study looked in the opposite direction and investigated the perspective of mental health practitioners on AAT with children and adolescents. A survey about AAT was completed by 300 mental health practitioners. Results indicated that 92% viewed AAT as a legitimate counseling modality, but only 12% had received training in AAT. However, nearly 60% of the practitioners were interested in receiving AAT education, training, and supervision (Hartwig & Smelser, 2018). This is extremely exciting for those of us actively incorporating AAT in our clinical practices. The more professionals who use AAT will hopefully result in AAT moving from being more of an exclusionary practice to an inclusionary practice within health care insurance policies. Additional studies of AAT for children with mental health disorders are needed. Researchers should strive to use higher tier methods and control as many variables as possible, so the results can be generalized and used to validate AAT for this population.

Autism spectrum disorders

Children with ASDs are excellent candidates for AAT because they may have difficulty developing social relationships and communicating with others. These two areas may improve because of the human—animal relationship bond that can occur between the therapy dog and the child. Most of the studies of AAT with children on the autism spectrum have focused on social participation, social-emotional abilities, physical function, attention, and communication. One study conducted in Spain targeted 3- and four-year-old preschoolers with ASD using a within-subject quasiexperimental study design. The children participated in nine 20-minute AAT sessions and were evaluated with different formal observation and survey measures before and after the AAT program. At the end of the study, the preschoolers showed improvements in social interaction and communication skills with a large effect size. Statistically significant improvements were found across variables such as child—therapy dog social relationships (e.g., pet the dog, talk to the dog, play a game with the dog) and child—therapist relationship (e.g., talking with the therapist, interacting with the therapist). The authors concluded that AAT enhances the social communication abilities of young children with ASD and should be considered a beneficial nonpharmacological therapy that can be integrated with traditional treatment (Ávila-Álvarez et al., 2020). Additional studies also found that AAT facilitated the development of social-emotional and social-communication abilities of children with autism. The children were able to understand and respond to the emotions of others during social interaction and use verbal language or an augmentative/alternative communication device to initiate and respond to communication with other people. Children with ASD also showed improvement following AAT in physical skills such as walking, climbing stairs, and balance with a decrease in risk of falls (Gómez-Calcerrada et al., 2021; Grigore & Bazgan, 2017). Multiple studies that investigated the parent's perspective of AAT programs for their children with ASD reported the majority of parents stated their children benefitted greatly from the AAT services. Specifically, the parents reported that participation in AAT facilitated their children's engagement, enjoyment, and motivation for life. These new abilities were observed in the children's communication with the therapy dog and other people; the child's ability to regulate their behavior, and the child's ability to participate in social situations (Kregiel et al., 2019; London et al., 2020).

A systematic review and a metaanalysis of AAT for children with AAT were published in 2021. The review targeted studies published between 2010 and 2019 and found 11 articles that met the inclusionary criteria. Results of these studies concluded that AAT had a positive effect on the targeted skills in children with autism following AAT in all but one study. It was noted that many of these peer-reviewed research studies did not adequately describe the methods and as such, so are unable to be replicated. Specifically, characteristics about the therapy animal, characteristics and baseline level of function for the participants with ASD, and descriptions of the procedures of the AAT intervention, all needed clarity and specificity. If this occurs on future studies, then reproducibility is possible and the literature foundation for AAT is made stronger (Çetin & Çuhadar, 2021). The metaanalysis study started with 1447 articles following a literature search, but only 16 articles met the inclusion criteria. Analysis of these 16 articles revealed improvements in social interaction, communication, and reduction of ASD characteristics in children with ASD following participation in AAT intervention. In addition, there was no correlation between dosage of AAT and effect size of the results. Conclusions of the metaanalysis indicate that AAT appears to benefit children with ASD, especially in the areas of social interaction and communication. It is interesting that both the systematic review and the metaanalysis reported that a majority of studies found used weak research designs. Future studies of AAT for children with ASD need to have clearer, more specific discussions of the methods, including child and dog participant characteristics and reproducible AAT procedures (Çetin & Çuhadar, 2021; Dimolareva & Dunn, 2021).

Severe disabilities

Children with severe cognitive, physical, and/or psychological disabilities have also participated in AAT with therapy dogs. These studies have targeted disruptive behaviors, fear and anxiety, communication and relationships with others, attention and hyperactivity, and activities of daily living. Overall results indicated AAT facilitated improvement in all of the above variables to different degrees (Andżelina Wolan-Nieroda et al., 2021;

Grabowska & Ostrowska, 2018; Hill et al., 2020; Juríčková et al., 2020; Lobato Rincón et al., 2021; Machová et al., 2018; Schuck et al., 2018; Vidal et al., 2020; Vodka et al., 2018). In one study, 14 children aged 3–12 years with severe cognitive and physical disabilities participated in 12 sessions of AAT. The AAT sessions were goal-oriented, planned, and structured therapy sessions conducted by a teacher and a psychologist. The therapy goals targeted facilitation of developmental milestones in the areas of psychomotor, social, cognitive, and communication skills. Analysis of the behavioral observations by two raters during each of the sessions with each child revealed improvements in all four variables at the end of the AAT intervention—postural control, language and communication, autonomy, and confidence. It should be noted that this study was terminated early due to the COVID-19 shut down in March 2020. Therefore, it is possible that even greater improvements could have been made with double the number of AAT sessions (Lobato Rincón et al., 2021).

Another study investigated the effects of AAT on 39 children with significant disabilities aged 4–9 years. The children participated in 30 minute once weekly AAT sessions from less than one year to three years. Three surveys were used—a diagnostic survey of developmental abilities, a parent questionnaire, and a survey developed specifically for this study that targeted AAT. Results of the surveys indicated that 92% of parents reported improvement in their child's motor abilities, 63% in communication abilities, 84% in balance and communication, 94% in increased attention to task, 82% in emotional functioning and expressing feelings and emotions, 72% in social skills, and 33% in reduction of aggression. Additional findings indicated that parents reported their children had better mood, motivation, and attitude toward attending and participating in their habilitative therapies. Overall, parents reported improvements in all three major areas of their child's life—physical, intellectual, and emotional—as a result of participation in AAT in addition to their regular long-term therapies (Grabowska & Ostrowska, 2018).

Children with fetal alcohol syndrome (FAS), communication impairment, and severe ASD participated in three different randomized control studies of the effects of AAT. The first of these studies evaluated the effectiveness of AAT with 33 children and adolescents with FAS who were randomly assigned to AAT or no AAT. Seventeen children participated in weekly, 45-minute AAT sessions for 12 weeks. Six AAT sessions were individual and six sessions occurred with three to four children in a group. The AAT sessions were conducted by two therapy dog teams and a psychologist. The psychologist developed the AAT sessions to target the treatment goals of the participating children. The second group of 16 children received treatment as usual for FAS without participation in AAT. Three parental surveys were completed at the initiation of AAT sessions and at the end of AAT sessions. The control group completed the measures at 12-week interval. The parent surveys targeted child behavior, social skills, and a global impression scale for severity and were completed with blind rater who did not participate in any other portion of the study. Results of the surveys were analyzed statistically using ANOVA, independent t-tests, χ 2, and Cohen's d tests. Results indicated significant improvements in social skills in the AAT group as compared to the control group with a large effect size. The severity of FAS symptoms decreased significantly in the AAT group with a medium effect size. There were no differences between the AAT and control groups on the variables of reduction of internalizing and externalizing symptoms (Vidal et al., 2020).

The second randomized controlled trial investigated the effects of AAT on communication in 69 children aged four to seven years of age with developmental dysphagia/severe developmental language disorder. The children were randomly assigned to either the experimental group who received traditional speech therapy with AAT or the control group who received traditional speech therapy without AAT. Measures of facial movement/expression and motor proficiency/movement for speech were given to the experimental group at the beginning of AAT (time 1) and end of AAT (time 2) and at the same time interval for the control group. Statistical analyses revealed significant differences between the two groups at time 2 for both measures. Specifically, improvements in facial expression and motor speech tasks were observed in the children who received speech therapy with AAT. The improved motor abilities included smiling, filling cheeks with air, blowing, and opening and closing their eyes. Therefore, AAT with therapy dogs may facilitate or enhance the effects of speech therapy. One benefit of incorporating AAT into speech therapy sessions is the increased motivation of the child to communicate with the therapy dog because of their relationship and bond (Machová et al., 2018). See Fig. 9.1.

The third randomized controlled trial investigated the effects of AAT on 22 children with ASD who were receiving occupational therapy services. The children, aged four to six years, were randomly placed into a treatment group or a waitlist group. The treatment sessions occurred weekly for 60 min for seven weeks. Session goals and activities were developed by an occupational therapist who also conducted the treatment with a certified

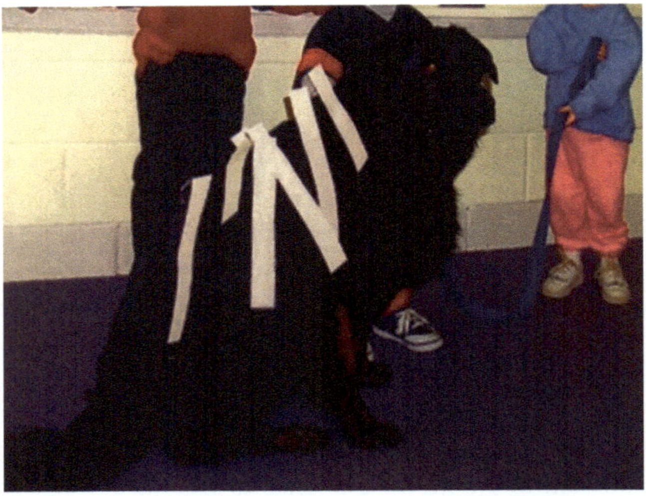

Figure 9.1 ***Black dog white stripes.*** Gabriel, a pet partner certified therapy dog, participated in speech therapy session sessions with preschoolers with language disorders. In this photo, the children "taped" white stripes on Gabriel so he would be "the same as a zebra." The children then took the white stripes off Gabriel so he would be "different than a zebra." *(Photograph by Beth Macauley. Written copyright permission on file. Permission for use of photograph on file.)*

therapy dog. Examples of treatment activities with and without AAT are provided in the article. All treatment sessions were videotaped and then the first and last session videotapes were coded by two nonstudy personnel who were blinded to the goals of the study. Results indicated improvement in time on task and attainment of goals following the seven weeks of OT integrating AAT, but the changes were not statistically significant. It is possible that conducting more treatment sessions or adding more participants could have added enough power to achieve statistical significance (Hill et al., 2020).

Reading disorders

One of the most popular and well-known applications of AAT is the addition of therapy dogs into the reading practice of children. This activity has occurred for over 25 years, but the first "official" program was established in 1999 in Utah. The Intermountain Therapy Animals organization founded the Reading Education Assistance Dogs (READ) program in which therapy dogs were paired with children learning to read. The group hypothesized that reading to dogs provided a nonjudgmental and safe environment which increased the child's motivation to read. Numerous studies were then initiated to document the expected outcomes of reading to dogs programs. A systematic review of the literature conducted in 2016 found 48 articles related to reading with dogs that met the inclusionary criteria. Extensive analyses of these 48 articles revealed that reading to dogs is beneficial to children and may assist with improvement in reading abilities. However, the quality of evidence from these studies is low because the majority used weak methodology, had a small number of participants, and/or did not use standardized measures (Hall et al., 2016).

Related studies from 2016 to 2022 found that the children experienced increased confidence and relaxation, lowered blood pressure, and increased reading abilities following participation in a reading to therapy dogs program (Barber & Proops, 2019; Jalongo, 2005; Jalongo et al., 2004; Juríčková et al., 2020; Rousseau & Tardif-Williams, 2019; Schretzmayer et al., 2017). A few of these studied targeted children in special education and children with ASD (Fung, 2017; Uccheddu et al., 2019). One study described the development of a reading to dogs program by an interprofessional group of teachers, therapy dog teams, and a psychologist (Steel et al., 2022).

Horses

Horses are the second most prevalent animal incorporated into AAT sessions with children. One difference between dogs and horses is that therapy certification programs for dogs exist, while therapy certification programs for horses do not exist. However, the Professional Association of Therapeutic Horsemanship, Intl. (PATH, Intl.) has certification standards for equestrian facilities that offer AAT with horses. These standards comprise written policies and procedures for how the horses are trained, prepared, and desensitized for therapy, maintenance of the horse's health and welfare, and policies and procedures for the retirement of a therapy horse. It is highly recommended that anyone who incorporates horses into AAT be aware of and meet these standards of the profession (PATH, Intl. Standards Manual, 2021).

Another difference between dogs and horses in AAT is use of terminology. Use of the term "AAT" assumes that dogs were the participating animal. Whereas when horses are included in AAT, terms such as equine therapy, equine-assisted therapy, equine-facilitated therapy, equine-facilitated mental health, equine learning, therapeutic riding, and hippotherapy have all been used in the literature. In order to clarify terminology, a task force with representatives from the Professional Association of

Therapeutic Horsemanship, Intl. (PATH, Intl.) and the American Hippotherapy Association, Inc. (AHA, Inc.) created a consensus document on terminology. The task force defined 3 categories of services that incorporate horses to benefit adults and children: therapy, learning, and horsemanship. The therapy category includes equine-assisted speech-language pathology, physical therapy, occupational therapy, counseling, and psychotherapy. The learning category includes equine-assisted learning in education and personal development. The horsemanship category includes adaptive riding/therapeutic riding and interactive vaulting. The American Hippotherapy Association encouraged practitioners to go above and beyond this document and use very clear and specific terminology such as "physical therapy incorporating horses (PTIH)" and "speech-language pathology incorporating hippotherapy" rather than "equine-assisted speech therapy" to help the public and health care agencies better understand that the treatment by licensed, credentialed therapists is paramount. The therapists incorporate horses as a treatment strategy used within their scope of practice (Macauley, 2022; Wood et al., 2021). Therefore, the overarching term Animal-Assisted Therapy Incorporating Horses (AATIH) will be used in this chapter, with subtypes according to profession such as Psychology Incorporating Horses (PIH), PTIH, and Social Work Incorporating Horses (SWIH).

Mental health

The incorporation of AATIH into psychological counseling and services for children with psychological and mental health disorders has increased greatly over the last 20 years. At-risk children, children with ADHD, PTSD, schizophrenia, mood disorders, depression, behavior disorders, children postsubstance, sexual, and physical abuse, and children post-traumatic experiences have all participated in psychological services incorporating AATIH. It is exciting that research has documented improvements in so many areas following AATIH for these children and youth.

Studies of the benefits of psychological services with horses for children utilized the therapy horses in different ways. These included riding, an "on the horse" activity, and ground work, grooming, and experiential activities which are "off the horse" activities. The experiential modality provides a unique learning experience for personal growth. Experiential learning involves developing a positive relationship between the child and the horse. As the positive relationship develops over time under the guidance of the treating therapist, the child or adolescent begins to develop positive emotions, empathy for others, a decrease in depression with increased hope for the future. Experiential learning typically occurs off the horse and is not a riding experience. Experiential programs as short as five weeks in length showed improvements in hope and quality of life in at-risk youth (Harvey et al., 2020; Mueller & McCullough, 2017; Sauer & Gill, 2020).

Children who experienced traumatic situations such as sexual abuse and parental substance abuse showed improvements in managing emotions such as decreasing depression and increasing positive mood and improving difficult behavior such as fighting. Studies also promoted the safe and secure nature of the AATIH environment which decreases stress and tension allowing the children to communicate more openly and have increased engagement and buy-in to the counseling activities (Dunlop & Tsantefski, 2018; Mueller & McCullough, 2017; Signal et al., 2017; Tsantefski et al., 2017). One study compared traditional counseling with equine-facilitated counseling for 30 children and youth ages 8–17 years. The psychometric measures were completed before the initiation of intervention, after six weeks of in-clinic counseling, and after 9–10 weeks of equine-facilitated counseling. Results indicated that while children showed improvements following both types of counseling, more improvements were observed following the equine-facilitated counseling. These results were consistent across age and sex of the participants. Conclusions state that equine-facilitated therapy is an effective therapeutic approach for children and adolescents following trauma including sexual abuse (Kemp et al., 2014).

A systematic review of the effects of equine-assisted activities and therapies in children and youth with ADHD was completed in 2021. The authors found nine studies through 2019 that met the inclusionary criteria. Results indicated that children with ADHD needed 15–40 minutes of time once or twice weekly over 4–32 weeks in order to reduce the symptoms associated with ADHD. Conclusions stated that while all studies documented that the children with ADHD benefitted from the equine-assisted activities and therapy, the high heterogeneity of variables combined with low methodological quality led to a lower level of evidence and support for the results (Pérez-Gómez et al., 2021).

Cerebral palsy

Children with cerebral palsy have participated in AATIH with great success. Improvements in gait and motor function (Antunes et al., 2016; Brady et al., 2021; Kwon et al., 2011; Manikowska et al., 2013; Matusiak-Wieczorek et al., 2020; McGibbon et al., 1998; Mutoh et al., 2018,

2019; Shurtleff & Engsberg, 2012), balance and sitting balance (Fisher-Pipher et al., 2017; Kang et al., 2012; Lakomy-Gawryszewska et al., 2017; Matusiak-Wieczorek et al., 2020), reaching behavior (Shurtleff et al., 2009), energy expenditure (McGibbon et al., 1998), articulation and respiration for speech (Macauley, 2007, 2021, 2022), and quality of life (Czerw, 2017; Silkwood-Sherer & McGibbon, 2022) have all been documented following AATIH. Within these studies, hippotherapy was the primary AATIH treatment strategy used and a within subject pretest posttest design was the primary methodology. Hippotherapy is described as the purposeful manipulation of equine movement as a therapy tool to engage sensory, neuromotor, and cognitive systems to promote functional outcomes (Pantera et al., 2015). During hippotherapy, the child sits on the horse's back and accommodates the three-dimensional movement of the horse's walk. This integration of movement between the horse and the child stimulates the child's motor and sensory systems in a systematic and repetitive manner resulting in improved neurological function. The treating therapist can then teach new skills and facilitate improvements to help the child reach their therapy goals. Hippotherapy is primarily used within physical therapy, occupational therapy, and speech-language pathology. The history of hippotherapy, its applications to skilled treatment within speech-language pathology, and clinical case reports of SLPIH were discussed in an article by Macauley (2022).

A systemic review of hippotherapy for children with cerebral palsy found 16 studies that met the inclusion criteria. Results of these studies showed improvements in motor function, posture, balance, motor coordination, lumbopelvic flexibility, and walking as well as symmetry of muscle contraction, reduction of spasticity, and better quality of life (Pantera et al., 2015).

From 2016 to 2022, additional studies have been published on the effects of hippotherapy for children with cerebral palsy. One particular study investigated the effects of hippotherapy on the three areas of the World Health Organization's International Classification of Functioning, Disability and Health for Children and Youth (ICF-CY) model in children with cerebral palsy. An ABA design was used with a 12-week baseline, 12-week hippotherapy intervention, and 12-week withdrawal. Body functions, activities, and life participation information were obtained through parent interview at the initial interview and at the end of each phase of the study. Fourteen children with cerebral palsy aged 3–8 years participated. Results indicated that significant improvements in all 3 ICF-CY areas including musculoskeletal and movement related functions, mobility of joint functions, muscle tone functions, involuntary movement reaction functions, major life areas, and play. These effects of hippotherapy were distinct from the Gross Motor Function Classification System (GMFCS) levels, and most of the improvements were observed in children with GMFCS levels I—III (Hsieh et al., 2017). Another study investigated the effects of hippotherapy on gait in children and adolescents with spastic cerebral palsy over 12 months. Gait and balance abilities were examined in 20 children with cerebral palsy aged 4—19 years at the beginning of hippotherapy and then at 12, 24, 36, and 48 weeks of hippotherapy intervention. Results indicated that the participants had increased their stride length and acceleration already at 12 weeks and walking speed, mean acceleration, and decreased horizontal/vertical displacement were observed at 48 weeks. In addition, all participants had improved scores on the Gross Motor Function Measure at the one-year mark (Mutoh et al., 2018).

A main criticism of many hippotherapy studies is that observational measures were used rather than objective or empirical. One study that directly addresses this criticism was conducted by Li et al. in 2021. Li and colleagues used distributed magnetic, angular rate, and gravity sensors on the body to evaluate the gait and gross motor function of children with cerebral palsy. Data were obtained at different time periods across one year of hippotherapy intervention. The data from the body sensors were analyzed using a multisensor data fusion algorithm. Results of these kinematic analyses revealed issues in limb stiffness, poor joint range of motion, scissors gait, and knee flexion gait were relieved throughout the hippotherapy intervention with the most change coming at the end of the study. Conclusions state that these results provide an empirical basis for incorporating hippotherapy in the rehabilitation of children with cerebral palsy (Li et al., 2021).

Hippotherapy has been used as a treatment strategy for children with cerebral palsy primarily during physical therapy treatment sessions followed by occupational therapy treatment sessions, and then speech-language pathology treatment sessions. A series of studies and presentations by Macauley focused on the effects of hippotherapy on articulation ability and respiratory control for speech in children with cerebral palsy. Diadochokinesis and respiratory measurements were obtained before and after five to seven speech-language pathology sessions incorporating hippotherapy. Results indicated that the children's ability to coordinate their articulators and their respiratory system for speech improved at the end of every hippotherapy session but returned to baseline at the beginning of the next session. Measurements taken at the beginning of the session showed change at the beginning of session seven. It is possible that longer intervention would have revealed more improvements across sessions (Macauley, 2007, 2021, 2022).

Autism spectrum disorders

Children and youth with ASDs respond well to AATIH. The main areas targeted in treatment with this population are relationship, social-emotional skills, and communication. The ability of the child with ASD to build a relationship with the therapy horse adds another dimension to the AATIH treatment sessions. A systematic review of equine-assisted interventions for children with ASD found 33 studies that met the inclusion criteria published between 1980 and 2015. Even though the literature review went back to 1980, the earliest study found was published in 2003. Since 2023, research articles were published in 12 different countries with different types of AATIH including therapeutic riding, hippotherapy, psychoeducational horseback riding, equine-facilitated learning, and riding for the disabled. Results of the 33 studies documented improvements in behavior such as reduction of maladaptive behaviors including emotional displays, stereotyped movements, irritability, hyperactivity, and aggression; interpersonal interactions such as social skills including improved relationships with family and friends; communication abilities such receptive and expressive communication both verbal and nonverbal; interpersonal interaction such as adaptive social behaviors, mood, and tone of parent—child interactions, social cognition, social communication, and overall social functioning. Therefore, although the cumulation of results documented the benefits of AATIH for children with autism across many different measures, the type, dosage, and providers of the intervention as well as activities that occurred during the AATIH sessions were very heterogeneous in nature (McDaniel Peters & Wood, 2017). Fig. 9.2.

Studies also showed that parents and caregivers reported that their children with ASD benefitted greatly from AATIH intervention. Parent's perspectives were recorded through surveys and structured interviews across studies. Themes that came out of these methods revealed that parents expressed that their children improved in self-concept, emotional well-being, improved self-regulatory behaviors, social abilities and communication, self-care abilities, and quality of life (Scotland-Coogan et al., 2021; Tan & Simmonds, 2018).

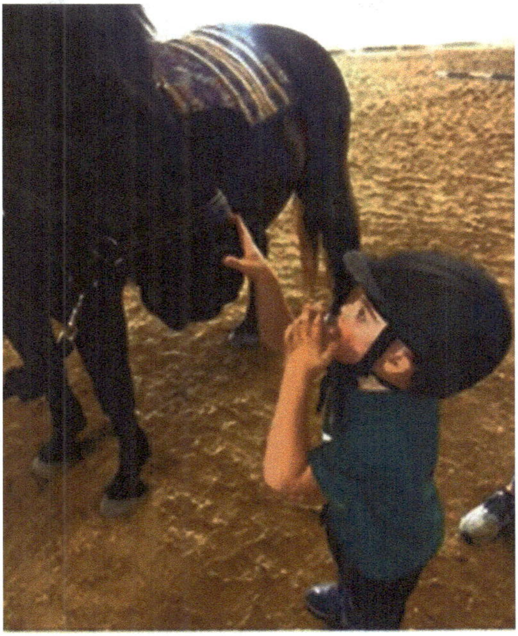

Figure 9.2 Horse and child. relationship through eye contact and touch between Babe, the therapy horse, and her young client after two speech-language pathology treatment sessions incorporating horses. *(Photograph taken by Beth Macauley. Written copyright permission on file. Permission to use the photograph on file.)*

Other animals

AAT can be accomplished with animals other than dogs and horses as long as those other animals meet the standard for use in AAT. These animals have included cats, guinea pigs, and elephants.

A study involving cats was completed by Boyer and Mundschenk (2014). They investigated whether AAT using cats was effective in promoting social communication between children with language impairments and typically developing peers by comparing the use of a live cat with a toy cat and a preferred activity during speech therapy sessions. Two of the three children who participated demonstrated more sustained social interaction when the live cat was included over the toy cat or the preferred activity. While two out of three is a majority, the low number of participants limits the generalizability of the results (Boyer & Mundschenk, 2014).

The study involving guinea pigs was completed by O'Haire et al. (2013). They investigated the interactions of children with ASDs with an adult and typically developing peers in the presence of two guinea pigs compared to toys. Ninety-nine children with ASD participated in the study. They were videotaped during three 10-min, free-play sessions with toys and three 10-min, free-play sessions with two guinea pigs. Two blinded observers coded the behavior of children with ASD and their peers. Participants with ASD demonstrated more social approach behaviors such as talking, looking at faces, and making tactile contact and received more social approaches from their peers in the presence of the guinea pigs than in the presence of toys. They also displayed more prosocial behaviors and positive affect such as smiling and laughing as well as less self-focused behaviors and negative affect such as frowning and crying in the presence of the guinea pigs when compared to toys. The researchers conclude that the presence of guinea pigs can significantly increase positive social behaviors among children with ASD (O'Haire, 2013).

Two studies incorporated elephants as therapy animals. Both studies were conducted in Thailand in 2014 and 2016 through the Thai Elephant-Assisted Therapy Program (TEATP). Sasithorn (2014) investigated the effects of the TEATP on social and maladaptive behaviors in children and youth with ASDs during either a 3-week or 6-week program. Results indicated that social participation significantly increased and maladaptive behaviors significantly decreased following both the short and the longer elephant-assisted programs. Satiansukpong (2016) investigated the effects of the TEATP on balance, postural control, and visual motor integration in children with Down syndrome. The participants were divided voluntarily into two groups: control and experimental. Both groups received regular school activities, but the experimental group added TEATP twice a week for 2 months. The results showed no significant difference in balance or postural control in either group of children, but the children who participated in the elephant-assisted therapy program had significant improvement in visual motor integration (Sasithorn et al., 2014; Satiansukpong et al., 2016).

Using elephants as therapy animals in Thailand is similar to using horses as therapy animals in Mexico. Both countries have domesticated and trained these animals for human use, and the elephants and horses have been integrated into the culture of the country. However, there are ethical questions about the use of elephants that include whether standards and policies for the use of elephants in AAT sessions has been written and is being followed, who is in charge of the elephant and can decide whether an elephant is stressed by their participation or not. It is hoped that people incorporating elephants into AAT sessions have written documentation for their use and are being mindful of the welfare of the elephants.

Future studies incorporating uncommon animals in AAT should address the ethical issues of that particular animal as well as discuss standards and policies for their use. As more and different animals are used as therapy animals, it is critical that people understand the pros and cons, the criteria, and the welfare of the animal and these issues should be included in published studies.

Conclusion

Children and animals, the combination is a powerful treatment strategy that is used within a plethora of helping professions. AAT with dogs and horses as well as cats, guinea pigs, and elephants have studies documenting their positive effects on preschoolers, children, and youth. Professional areas of psychology, psychiatry, mental health counseling, social work, physical therapy, occupational therapy, and speech-language pathology have incorporated therapy animals within their scope of practice to improve the lives of their young patients. Areas of benefit have included decreased stress and depression, improvement in motor control and walking, balance, reaching, socio-emotional regulation, receptive and expressive communication, articulation agility, management of maladaptive behaviors, increased engagement and participation in life activities, more positive mood, and improved quality of life. One of the main barriers to increasing the availability of AAT for children is the lack of education and training of health care and education professionals. Many surveyed professionals stated that they wanted to include AAT as a service but did not know how. Practitioners interested in AAT should contact the organization responsible for education and training for the area in which they are interested or contact a person within their profession who is offering AAT (Hartwig & Smelser, 2018). As more practitioners include AAT in their clinical services and the benefits of AAT for children are documented with stronger research studies, AAT will become more accepted as the state-of-the-art treatment strategy that it is. Come join the fun!

References

Acerboni, V., De Piero, M., Dedato, R., & Bernoni, M. (2013). Implementing pet therapy in the pediatric intensive care. *Scenario: Official Italian Journal of Aniarti, 30*(2), 21–22. Supplement https://ezproxy.gvsu.edu/login?url=https://search.ebscohost.com/login.aspx?direct=true&AuthType=ip,sso&db=edo&AN=91543497&site=eds-live&scope=site

Allen, B., Shenk, C. E., Dreschel, N. E., Wang, M., Bucher, A. M., Desir, M. P., Chen, M. J., & Grabowski, S. R. (2021). Integrating animal-assisted therapy into TF-CBT for abused youth with PTSD: A randomized controlled feasibility trial. *Child Maltreatment*. https://doi.org/10.1177/1077559520988790

Antunes, F. N., Pinho, A. S., do Kleiner, A. F. R., Salazar, A. P., Eltz, G. D., de Oliveira Junior, A. A., Cechetti, F., Galli, M., & Pagnussat, A. S. (2016). Different horse's paces during hippotherapy on spatio-temporal parameters of gait in children with bilateral spastic cerebral palsy: A feasibility study. *Research in Developmental Disabilities, 59*, 65–72. https://doi.org/10.1016/j.ridd.2016.07.015

Ávila-Álvarez, A., Alonso-Bidegain, M., De-Rosende-Celeiro, I., Vizcaíno-Cela, M., Larrañeta-Alcalde, L., & Torres-Tobío, G. (2020). Improving social participation of children with autism spectrum disorder: Pilot testing of an early animal-assisted intervention in Spain. *Health & Social Care in the Community, 28*(4), 1220–1229. https://ezproxy.gvsu.edu/login?url=https://search.ebscohost.com/login.aspx?direct=true&AuthType=ip,sso&db=c9h&AN=143548195&site=eds-live&scope=site

Balluerka, N., Muela, A., Amiano, N., & Caldentey, M. A. (2014). Influence of animal-assisted therapy (AAT) on the attachment representations of youth in residential care. *Children and Youth Services Review, 42*, 103–109. https://ezproxy.gvsu.edu/login?url=https://search.ebscohost.com/login.aspx?direct=true&AuthType=ip,sso&db=edselp&AN=S0190740914001571&site=eds-live&scope=site

Barber, O., & Proops, L. (2019). Low-ability secondary school students show emotional, motivational, and performance benefits when reading to a dog versus a teacher. *Anthrozoos, 32*(4), 503–518. https://ezproxy.gvsu.edu/login?url=https://search.ebscohost.com/login.aspx?direct=true&AuthType=ip,sso&db=asn&AN=137235810&site=eds-live&scope=site

Boyer, V., & Mundschenk, N. (2014). Using animal-assisted therapy to facilitate social communication: A pilot study. *Canadian Journal of Speech-Language Pathology & Audiology, 38*(1), 26–38.

Brady, H. A., James, C. R., Dendy, D. W., Irwin, T. A., Thompson, L. D., & Camp, T. M. (2021). Gross motor skills and gait performance in two- and three-year-old children with developmental delay participating in hippotherapy. *Journal of Equine Veterinary Science, 99*, 103359. https://doi.org/10.1016/j.jevs.2020.103359

Calcaterra, V., Veggiotti, P., Palestrini, C., De Giorgis, V., Raschetti, R., Tumminelli, M., Mencherini, S., Papotti, F., Klersy, C., Albertini, R., Ostuni, S., & Pelizzo, G. (2015). Post-operative benefits of animal-assisted therapy in pediatric surgery: A randomised study. *PLOS ONE, 10*(6). https://doi.org/10.1371/journal.pone.0125813

Çetin, D., & Çuhadar, S. (2021). A review of studies conducted with animal assisted interventions for children with autism spectrum disorder. *Otizm Spektrum Bozukluğu Olan Çocuklara Yönelik Hayvan Destekli Müdahalelerle Yürütülen Arastırmaların İncelenmesi., 13*(3), 619–639. https://ezproxy.gvsu.edu/login?url=https://search.ebscohost.com/login.aspx?direct=true&AuthType=ip,sso&db=e5h&AN=151243284&site=eds-live&scope=site

Czerw, M. (2017). Hippotherapy as a form to support social functioning of children with cerebral palsy. *International Multidisciplinary Scientific Conference on Social Sciences & Arts SGEM*, 255–261. https://doi.org/10.5593/sgemsocial2017/34

Dietz, T. J., Davis, D., & Pennings, J. (2012). Evaluating animal-assisted therapy in group treatment for child sexual abuse. *Journal of Child Sexual Abuse, 21*(6), 665–683. https://ezproxy.gvsu.edu/login?url=https://search.ebscohost.com/login.aspx?direct=true&AuthType=ip,sso&db=edshol&AN=edshol.hein.journals.wcsa21.52&site=eds-live&scope=site

Dimolareva, M., & Dunn, T. J. (2021). Animal-assisted interventions for school-aged children with autism spectrum disorder: A meta-analysis. *Journal of Autism & Developmental Disorders, 51*(7), 2436–2449. https://ezproxy.gvsu.edu/login?url=https://search.ebscohost.com/login.aspx?direct=true&AuthType=ip,sso&db=c9h&AN=150794135&site=eds-live&scope=site

Dunlop, K., & Tsantefski, M. (2018). A space of safety: Children's experience of equine-assisted group therapy. *Child & Family Social Work, 23*(1), 16–24. https://ezproxy.gvsu.edu/login?url=https://search.ebscohost.com/login.aspx?direct=true&AuthType=ip,sso&db=c9h&AN=127423952&site=eds-live&scope=site

Feng, Y., Lin, Y., Zhang, N., Jiang, X., & Zhang, L. (2021). Effects of animal-assisted therapy on hospitalized children and teenagers: A systematic review and meta-analysis. *Journal of Pediatric Nursing, 60*, 11–23. https://ezproxy.gvsu.edu/login?url=https://search.ebscohost.com/login.aspx?direct=true&AuthType=ip,sso&db=edo&AN=152739815&site=eds-live&scope=site

Fisher-Pipher, S., Kenyon, L. K., & Westman, M. (2017). Improving balance, mobility, and dual-task performance in an adolescent with cerebral palsy: A case report. *Physiotherapy Theory & Practice, 33*(7), 586–595. https://doi.org/10.1080/09593985.2017.1323359

Flynn, E., Roguski, J., Wolf, J., Trujillo, K., Tedeschi, P., & Morris, K. N. (2019). A randomized controlled trial of animal-assisted therapy as an adjunct to intensive family preservation services. *Child Maltreatment, 24*(2), 161–168. https://ezproxy.gvsu.edu/login?url=https://search.ebscohost.com/login.aspx?direct=true&AuthType=ip,sso&db=eue&AN=136749066&site=eds-live&scope=site

Fung, S. (2017). Canine-assisted reading programs for children with special educational needs: Rationale and recommendations for the use of dogs in assisting learning. *Educational Review, 69*(4), 435–450. https://ezproxy.gvsu.edu/login?url=https://search.ebscohost.com/login.aspx?direct=true&AuthType=ip,sso&db=asn&AN=123287654&site=eds-live&scope=site

Gilmer, M. J., Baudino, M., Goddard, A. T., Vickers, D., & Akard, T. (2016). Animal-assisted therapy in pediatric palliative care. *Nursing Clinics of North America, 51*(3), 381.

Gómez-Calcerrada, I., Lavín-Pérez, A. M., Villafaina, S., Rueda-Rubio, J. C., Rivera-Martín, B., González-García, I., Merellano-Navarro, E., Mateo-Garitagoitia, M., González-Plaza, C., & Collado-Mateo, D. (2021). Effects of dog-assisted therapy on the physical function and communication skills of adults with autism: A study protocol for a controlled study. *Applied Sciences, 11*(22), 10650, 2076-3417 https://ezproxy.gvsu.edu/login?url=https://search.ebscohost.com/login.aspx?direct=true&AuthType=ip,sso&db=aps&AN=153790210&site=eds-live&scope=site

Grabowska, I., & Ostrowska, B. (2018). Evaluation of the effectiveness of canine assisted therapy as a complementary method of rehabilitation in disabled children. *Physiotherapy Quarterly, 26*(3), 18–27. https://doi.org/10.5114/pq.2018.78376

Granger, B. P., Kogan, L., Fitchett, J., & Helmer, K. (1998). A human-animal intervention team Approach to animal-assisted therapy. *Anthrozoös, 11*(3), 172–176. https://doi.org/10.2752/089279398787000689

Grigore, A. N., & Bazgan, M. (2017). Effects of assisted animal therapy on the development of socio-emotional abilities of children with autism. *Bulletin of the Transilvania University of Brasov. Series VII: Social Sciences. Law, 10*, 231–238. https://ezproxy.gvsu.edu/login?url=https://search.ebscohost.com/login.aspx?direct=true&AuthType=ip,sso&db=e5h&AN=129478411&site=eds-live&scope=site

Hall, S. S., Gee, N. R., & Mills, D. S. (2016). Children reading to dogs: A systematic review of the literature. *PLoS ONE, 11*(2), 1–22. https://ezproxy.gvsu.edu/login?url=https://search.ebscohost.com/login.aspx?direct=true&AuthType=ip,sso&db=asn&AN=113218336&site=eds-live&scope=site

Hartwig, E., & Smelser, Q. (2018). Practitioner perspectives on animal-assisted counseling. *Journal of Mental Health Counseling, 40*(1), 43–57.

Harvey, C., Jedlicka, H., & Martinez, S. (2020). A program evaluation: Equine-assisted psychotherapy outcomes for children and adolescents. *Child & Adolescent Social Work Journal, 37*(6), 665–675. https://doi.org/10.1007/s10560-020-00705-0

Hill, J., Ziviani, J., Driscoll, C., Teoh, A. L., Chua, J. M., & Cawdell-Smith, J. (2020). Canine assisted occupational therapy for children on the autism spectrum: A pilot randomised control trial. *Journal of Autism & Developmental Disorders, 50*(11), 4106–4120. https://ezproxy.gvsu.edu/login?url=https://search.ebscohost.com/login.aspx?direct=true&AuthType=ip,sso&db=c9h&AN=146431894&site=eds-live&scope=site

Hinic, K., Kowalski, M. O., Holtzman, K., & Mobus, K. (2019). The effect of a pet therapy and comparison intervention on anxiety in hospitalized children. *Journal of Pediatric Nursing, 46*, 55–61. https://ezproxy.gvsu.edu/login?url=https://search.ebscohost.com/login.aspx?direct=true&AuthType=ip,sso&db=edselp&AN=S0882596318304573&site=eds-live&scope=site

Hsieh, Y.-L., Yang, C.-C., Sun, S.-H., Chan, S.-Y., Wang, T.-H., & Luo, H.-J. (2017). Effects of hippotherapy on body functions, activities and participation in children with cerebral palsy based on ICF-CY assessments. *Disability & Rehabilitation, 39*(17), 1703–1713. https://ezproxy.gvsu.edu/login?url=https://search.ebscohost.com/login.aspx?direct=true&AuthType=ip,sso&db=edb&AN=123149821&site=eds-live&scope=site

Ichitani, T., & Cunha, M. (2016). Effects of animal-assisted activity on self-reported feelings of pain in hospitalized children and adolescents. *Psicologia: Reflexão e Critica, 29*(1), 1–10. https://ezproxy.gvsu.edu/login?url=https://search.ebscohost.com/login.aspx?direct=true&AuthType=ip,sso&db=edb&AN=118584154&site=eds-live&scope=site

Jalongo, M. R. (2005). What are all these dogs doing at school?" Using therapy dogs to promote children's reading practice. *Childhood Education, 81*(3), 152. https://ezproxy.gvsu.edu/login?url=https://search.ebscohost.com/login.aspx?direct=true&AuthType=ip,sso&db=edsbro&AN=edsbro.A134311927&site=eds-live&scope=site

Jalongo, M. R., Astorino, T., & Bomboy, N. (2004). Canine visitors: The influence of therapy dogs on young children's learning and well-being in classrooms and hospitals. *Early Childhood Education Journal, 32*(1), 9–16.

Jennings, M. L., Granger, D. A., Bryce, C. I., Twitchell, D., Yeakel, K., & Teaford, P. A. (2021). Effect of animal assisted interactions on activity and stress response in children in acute care settings. *Comprehensive Psychoneuroendocrinology, 8*. https://ezproxy.gvsu.edu/login?url=https://search.ebscohost.com/login.aspx?direct=true&AuthType=ip,sso&db=edselp&AN=S2666497621000503&site=eds-live&scope=site

Jones, M. G., Rice, S. M., & Cotton, S. M. (2019). Incorporating animal-assisted therapy in mental health treatments for adolescents: A systematic review of canine assisted psychotherapy. *PLoS ONE, 14*(1), e0210761. https://doi.org/10.1371/journal.pone.0210761

Juríčková, V., Bozděchová, A., Machová, K., & Vadroňová, M. (2020). Effect of animal assisted education with a dog within children with ADHD in the classroom: A case study. *Child and Adolescent Social Work Journal, 37*(6), 677–684. https://doi.org/10.1007/s10560-020-00716-x

Kang, H., Jung, J., & Yu, J. (2012). Effects of hippotherapy on the sitting balance of children with cerebral palsy: A randomized control trial. *Journal of Physical Therapy Science, 24*(9), 833–836. https://doi.org/10.1589/jpts.24.833

Kelly, M. A., & Cozzolino, C. A. (2015). Helping at-risk youth overcome trauma and substance abuse through animal-assisted therapy. *Contemporary Justice Review, 18*(4), 421–434. https://ezproxy.gvsu.edu/login?url=https://search.ebscohost.com/login.aspx?direct=true&AuthType=ip,sso&db=edb&AN=111728137&site=eds-live&scope=site

Kemp, K., Signal, T., Botros, H., Taylor, N., & Prentice, K. (2014). Equine facilitated therapy with children and adolescents who have been sexually abused: A program evaluation study. *Journal of Child & Family Studies, 23*(3), 558–566. https://ezproxy.gvsu.edu/login?url=https://search.ebscohost.com/login.aspx?direct=true&AuthType=ip,sso&db=pbh&AN=94763077&site=eds-live&scope=site

Kregiel, A., Zaworski, K., & Kołodziej, E. (2019). Effects of animal-assisted therapy on parent-reported behaviour and motor activity of children with autism spectrum disorder. *Health Problems of Civilization, 13*(4), 273–278. https://doi.org/10.5114/hpc.2019.90013

Kwon, J.-Y., Chang, H. J., Lee, J. Y., & Lee, P. K. (2011). Effects of hippotherapy on gait parameters in children with bilateral spastic cerebral palsy. *Archives of Physical Medicine and Rehabilitation, 92*(5), 774–779. https://doi.org/10.1016/j.apmr.2010.11.031

Lakomy-Gawryszewska, A. A., Józefowicz, K., Raniszewska, A., Langer, D., Hansdorfer-Korzon, R., Bieszczad, D., Górska, K., Cichoń-Kotek, M., & Pilarska, E. (2017). The impact of hippotherapy on the quality of trunk stabilisation, evaluated by EMG biofeedback, in children with infantile cerebral palsy. *Polish Annals of Medicine, 24*(1), 9–12. https://doi.org/10.1016/j.poamed.2016.06.001

Li, J., Wang, Z., Qiu, S., Zhao, H., Wang, J., Shi, X., Liang, B., & Fortino, G. (2021). Multi-body sensor data fusion to evaluate the hippotherapy for motor ability improvement in children with cerebral palsy. *Information Fusion, 70*, 115–128. https://doi.org/10.1016/j.inffus.2021.01.002

Lobato Rincón, L. L., Rivera Martín, B., Medina Sánchez, M.Á., Villafaina, S., Merellano-Navarro, E., & Collado-Mateo, D. (2021). Effects of dog-assisted education on physical and communicative skills in children with severe and Multiple disabilities: A pilot study. *Animals, 11*(6), 1741, 2076-2615 https://ezproxy.gvsu.edu/login?url=https://search.ebscohost.com/login.aspx?direct=true&AuthType=ip,sso&db=edb&AN=151085176&site=eds-live&scope=site

London, M. D., Mackenzie, L., Lovarini, M., Dickson, C., & Alvarez-Campos, A. (2020). Animal assisted therapy for children and adolescents with autism spectrum disorder: Parent perspectives. *Journal of Autism & Developmental Disorders, 50*(12), 4492–4503. https://ezproxy.gvsu.edu/login?url=https://search.ebscohost.com/login.aspx?direct=true&AuthType=ip,sso&db=eft&AN=147103404&site=eds-live&scope=site

Macauley, B. L. (2007). The effects of hippotherapy on respiration and motor speech in children with cerebral palsy. *Scientific and Educational Journal of Therapeutic Riding*, 4–11.

Macauley, B. L. (2021). Animal-assisted intervention in speech-language pathology. In *Animal-assisted interventions for health and human service professionals* (pp. 213–245). Nova Science Publishers.

Macauley, B. L. (2022). speech-language pathology incorporating horses. *Seminars in Speech & Language, 43*(1), 35–53. https://doi.org/10.1055/s-0041-1741553

Machová, K., Kejdanová, P., Svobodová, I., Bajtlerová, I., Procházková, R., & Mezian, K. (2018). Canine-assisted speech therapy for children with communication impairments: A randomized controlled trial. *Anthrozoos, 31*(5), 587–598. https://ezproxy.gvsu.edu/login?url=https://search.ebscohost.com/login.aspx?direct=true&AuthType=ip,sso&db=pbh&AN=131891571&site=eds-live&scope=site

Manikowska, F., Jóźwiak, M., Idzior, M., Chen, P.-J. B., & Tarnowski, D. (2013). The effect of a hippotherapy session on spatiotemporal parameters of gait in children with cerebral palsy - pilot study. *Ortopedia, Traumatologia, Rehabilitacja, 15*(3), 253. http://gvsu.summon.serialssolutions.com.ezproxy.gvsu.edu/2.0.0/link/0/eLvHCXMwY2AwNtIz0EUrE5LMjMxTQZOHaYZJKcBOF7BVYQiqbdPMkpOAPQYjlIhHKuDdhBiYUvNEGeTcXEOcPXQhdUR8AeQghnhgdWMJavAYiTHwJoIWhOeVgDeOpUgwKAAbzikGlkapSZbmSSYmiaYWyYaJZgappuYpaebAZnSqJIM4DtMA5x8yGQ

Matusiak-Wieczorek, E., Dziankowska-Zaborszczyk, E., Synder, M., & Borowski, A. (2020). The influence of hippotherapy on the body posture in a sitting position among children with cerebral palsy. *International Journal of Environmental Research and Public Health, 17*(18). https://doi.org/10.3390/ijerph17186846

McDaniel Peters, B., & Wood, W. (2017). Autism and equine-assisted interventions: A systematic mapping review. *Journal of Autism & Developmental Disorders, 47*(10), 3220–3242. https://doi.org/10.1007/s10803-017-3219-9

McGibbon, N. H., Andrade, C., Widener, G., Cintas, H. L., McGibbon, N. H., Andrade, C. K., Widener, G., & Cintas, H. L. (1998). Effect of an equine-movement therapy program on gait, energy expenditure, and motor function in children with spastic cerebral palsy: A pilot study. *Developmental Medicine & Child Neurology, 40*(11), 754–762. https://ezproxy.gvsu.edu/login?url=https://search.ebscohost.com/login.aspx?direct=true&AuthType=ip,sso&db=ccm&AN=107215715&site=eds-live&scope=site

Muela, A., Balluerka, N., Sansinenea, E., Machimbarrena, J. M., Garcia-Ormaza, J., Ibarretxe, N., Eguren, A., & Baigorri, P. (2021). A social-emotional learning program for suicide prevention through animal-assisted intervention. *Animals, 11*(12), 3375. https://doi.org/10.3390/ani11123375

Mueller, M., & McCullough, L. (2017). Effects of equine-facilitated psychotherapy on post-traumatic stress symptoms in youth. *Journal of Child & Family Studies, 26*(4), 1164–1172. https://ezproxy.gvsu.edu/login?url=https://search.ebscohost.com/login.aspx?direct=true&AuthType=ip,sso&db=ssf&AN=121699976&site=eds-live&scope=site

Mutoh, T., Mutoh, T., Tsubone, H., Takada, M., Doumura, M., Ihara, M., Shimomura, H., Taki, Y., & Ihara, M. (2018). Impact of serial gait analyses on long-term outcome of hippotherapy in children and adolescents with cerebral palsy. *Complementary Therapies in Clinical Practice, 30*, 19–23. https://ezproxy.gvsu.edu/login?url=https://search.ebscohost.com/login.aspx?direct=true&AuthType=ip,sso&db=edselp&AN=S1744388117304085&site=eds-live&scope=site

Mutoh, T., Mutoh, T., Tsubone, H., Takada, M., Doumura, M., Ihara, M., Shimomura, H., Taki, Y., & Ihara, M. (2019). Effect of hippotherapy on gait symmetry in children with cerebral palsy: A pilot study. *Clinical & Experimental Pharmacology & Physiology, 46*(5), 506–509. https://doi.org/10.1111/1440-1681.13076

Nahm, N., Lubin, J., Lubin, J., Bankwitz, B. K., Castelaz, M., Chen, X., Shackson, J. C., Aggarwal, M. N., & Totten, V. Y. (2012). Therapy dogs in the emergency department. *Western Journal of Emergency Medicine: Integrating Emergency Care with Population Health, 13*(4), 363–365. https://ezproxy.gvsu.edu/login?url=https://search.ebscohost.com/login.aspx?direct=true&AuthType=ip,sso&db=ccm&AN=107990263&site=eds-live&scope=site

Nilsson, M. L., Funkquist, E.-L., Edner, A., & Engvall, G. (2019). Children report positive experiences of animal-assisted therapy in paediatric hospital care. *ACTA PAEDIATRICA*, 1049–1056. https://doi.org/10.1111/apa.15047

O'Haire, M. (2013). Animal-assisted intervention for autism spectrum disorder: A systematic literature review. *Journal of Autism & Developmental Disorders, 43*(7), 1606–1622. https://ezproxy.gvsu.edu/login?url=https://search.ebscohost.com/login.aspx?direct=true&AuthType=ip,sso&db=eft&AN=88227432&site=eds-live&scope=site

Pantera, E., Vernay, D., Gautheron, V., & Laffont, I. (2015). Does hippotherapy improve motor function in children with cerebral palsy? Systematic review. *Annals of Physical and Rehabilitation Medicine, 58*, e131. https://doi.org/10.1016/j.rehab.2015.07.311

PATH, Intl. Standards Manual. (2021). www.pathintl.org/path-intl-membership/my-path-intl-membership/standards-manual

Perez, M., Cuscaden, C., Somers, J. F., Simms, N., Shaheed, S., Kehoe, L. A., Holowka, S. A., Aziza, A. A., Shroff, M. M., & Greer, M.-L. C. (2019). Easing anxiety in preparation for pediatric magnetic resonance imaging: A pilot study using animal-assisted therapy. *Pediatric Radiology, 49*(8), 1000–1009. https://doi.org/10.1007/s00247-019-04407-3

Pérez-Gómez, J., Amigo-Gamero, H., Collado-Mateo, D., Barrios-Fernandez, S., Muñoz-Bermejo, L., Garcia-Gordillo, M.Á., Carlos-Vivas, J., & Adsuar, J. C. (2021). Equine-assisted activities and therapies in children with attention-deficit/hyperactivity disorder: A systematic review. *Journal of Psychiatric & Mental Health Nursing, 28*(6), 1079–1091. John Wiley & Sons, Inc. https://ezproxy.gvsu.edu/login?url=https://search.ebscohost.com/login.aspx?direct=true&AuthType=ip,sso&db=c9h&AN=153608393&site=eds-live&scope=site

Prothmann, A., Bienert, M., & Ettrich, C. (2006). Dogs in child psychotherapy: Effects on state of mind. *Anthrozoos, 19*(3), 265–277. https://ezproxy.gvsu.edu/login?url=https://search.ebscohost.com/login.aspx?direct=true&AuthType=ip,sso&db=pbh&AN=23416724&site=eds-live&scope=site

Reide, D. (1987). The relationship between man and horse with reference to medicine throughout the ages. In , *Vol. 52. People, animals, environment* (pp. 26–28).

Rousseau, C. X., & Tardif-Williams, C. Y. (2019). Turning the page for spot: The potential of therapy dogs to support reading motivation among young children. *Anthrozoos, 32*(5), 665–677. https://ezproxy.gvsu.edu/login?url=https://search.ebscohost.com/login.aspx?direct=true&AuthType=ip,sso&db=asn&AN=138734478&site=eds-live&scope=site

Sasithorn, S.-U., Pongsaksri, M., Sasat, D., Vittayakorn, S., & Satiansukpong, N. (2014). The effect of Thai elephant-assisted therapy program on social and maladaptive behaviors for children and adolescents with autistic spectrum disorder. *Journal of Associated Medical Sciences, 47*(3), 162–162 https://ezproxy.gvsu.edu/login?url=https://search.ebscohost.com/login.aspx?direct=true&AuthType=ip,sso&db=edsdoj&AN=edsdoj.6b527cde3d2d4d50a00a7ce5572357cc&site=eds-live&scope=site

Satiansukpong, N., Pongsaksri, M., & Sasat, D. (2016). Thai elephant-assisted therapy programme in children with down Syndrome. *Occupational Therapy International, 23*(2), 121–131. https://ezproxy.gvsu.edu/login?url=https://search.ebscohost.com/login.aspx?direct=true&AuthType=ip,sso&db=c9h&AN=115775565&site=eds-live&scope=site

Sauer, A. N. K., & Gill, C. S. (2020). Treating disruptive mood dysregulation disorder: An integrated adlerian and equine therapy approach. *Journal of Individual Psychology, 76*(4), 372–385. https://ezproxy.gvsu.edu/login?url=https://search.ebscohost.com/login.aspx?direct=true&AuthType=ip,sso&db=pbh&AN=147693552&site=eds-live&scope=site

Schretzmayer, L., Kotrschal, K., & Beetz, A. (2017). Minor immediate effects of a dog on children's reading performance and physiology. *Frontiers in Veterinary Science, 4*, 90. https://doi.org/10.3389/fvets.2017.00090

Schuck, S. E., Johnson, H. L., Abdullah, M. M., Stehli, A., Fine, A. H., & Lakes, K. D. (2018). The role of animal assisted intervention on improving self-esteem in children with attention deficit/hyperactivity disorder. *Frontiers in Pediatrics*, 6. https://doi.org/10.3389/fped.2018.00300

Scotland-Coogan, D., Whitworth, J. D., & O'Brien, C. N. (2021). Caregiver perceptions of the benefits of hippotherapy for children with various disorders, disabilities, and medical conditions. *Journal of Human Behavior in the Social Environment, 31*(8), 1054–1076. https://ezproxy.gvsu.edu/login?url=https://search.ebscohost.com/login.aspx?direct=true&AuthType=ip,sso&db=hsi&AN=152758940&site=eds-live&scope=site

Seivert, N. P., Cano, A., Casey, R. J., Johnson, A., & May, D. K. (2018). Animal assisted therapy for incarcerated youth: A randomized controlled trial. *Applied Developmental Science, 22*(2), 139–153. https://ezproxy.gvsu.edu/login?url=https://search.ebscohost.com/login.aspx?direct=true&AuthType=ip,sso&db=pbh&AN=128422063&site=eds-live&scope=site

Shurtleff, T., & Engsberg, J. (2012). Long-term effects of hippotherapy on one child with cerebral palsy: A research case study. *The British Journal of Occupational Therapy, 75*(8), 359–366. https://doi.org/10.4276/030802212X13433105374279

Shurtleff, T. L., Standeven, J. W., & Engsberg, J. R. (2009). Changes in dynamic trunk/head stability and functional reach after hippotherapy. *Archives of Physical Medicine and Rehabilitation, 90*(7), 1185–1195. https://doi.org/10.1016/j.apmr.2009.01.026

Signal, T., Taylor, N., Prentice, K., McDade, M., & Burke, K. J. (2017). Going to the dogs: A quasi-experimental assessment of animal assisted therapy for children who have experienced abuse. *Applied Developmental Science, 21*(2), 81–93. https://ezproxy.gvsu.edu/login?url=https://search.ebscohost.com/login.aspx?direct=true&AuthType=ip,sso&db=pbh&AN=122343096&site=eds-live&scope=site

Silkwood-Sherer, D. J., & McGibbon, N. H. (2022). Can hippotherapy make a difference in the quality of life of children with cerebral palsy? A pragmatic study. *Physiotherapy Theory And Practice*, 390–400. https://doi.org/10.1080/09593985.2020.1759167

Silva, N., & Osorio, F. (2000). Impact of an animal-assisted therapy programme on physiological and psychosocial variables of paediatric oncology patients. *PLoS ONE, 3*(4), e0194731.

Steel, J., Williams, J. M., & McGeown, S. (2022). Teacher-researcher collaboration in animal-assisted education: Co-Designing a reading to dogs intervention. *Educational Research,* 1–19. https://ezproxy.gvsu.edu/login?url=https://search.ebscohost.com/login.aspx?direct=true&AuthType=ip,sso&db=asn&AN=154779906&site=eds-live&scope=site

Stefanini, Cristina, Maria, Martino, A., Bacci, B., & Tani, F. (2016). The effect of animal-assisted therapy on emotional and behavioral symptoms in children and adolescents hospitalized for acute mental disorders. *European Journal of Integrative Medicine, 8*(2), 81–88. https://ezproxy.gvsu.edu/login?url=https://search.ebscohost.com/login.aspx?direct=true&AuthType=ip,sso&db=edselp&AN=S187638201630018X&site=eds-live&scope=site

Stefanini, M. C., Martino, A., Allori, P., Galeotti, F., & Tani, F. (2015). The use of animal-assisted therapy in adolescents with acute mental disorders: A randomized controlled study. *Complementary Therapies in Clinical Practice, 21*(1), 42–46. https://ezproxy.gvsu.edu/login?url=https://search.ebscohost.com/login.aspx?direct=true&AuthType=ip,sso&db=edselp&AN=S1744388115000031&site=eds-live&scope=site

Syzmanski, T., Casey, R. J., Johnson, A., Cano, A., Albright, D., & Seivert, N. P. (2018). Dog training intervention shows social-cognitive change in the Journals of incarcerated youth. *Frontiers In Veterinary Science, 5*, 302. https://doi.org/10.3389/fvets.2018.00302

Tan, V. X.-L., & Simmonds, J. G. (2018). Parent perceptions of psychosocial outcomes of equine-assisted interventions for children with autism spectrum disorder. *Journal of Autism & Developmental Disorders, 48*(3), 459–469. https://ezproxy.gvsu.edu/login?url=https://search.ebscohost.com/login.aspx?direct=true&AuthType=ip,sso&db=pbh&AN=128167420&site=eds-live&scope=site

Trujillo, K. C., Kuo, G. T., Hull, M. L., Ingram, A. E., & Thurstone, C. C. (2020). Engaging adolescents: Animal assisted therapy for adolescents with psychiatric and substance use disorders. *Journal of Child & Family Studies, 29*(2), 307–314. https://ezproxy.gvsu.edu/login?url=https://search.ebscohost.com/login.aspx?direct=true&AuthType=ip,sso&db=eft&AN=141340109&site=eds-live&scope=site

Tsantefski, M., Briggs, L., Griffiths, J., & Tidyman, A. (2017). An open trial of equine-assisted therapy for children exposed to problematic parental substance use. *Health & Social Care in the Community, 25*(3), 1247–1256. https://ezproxy.gvsu.edu/login?url=https://search.ebscohost.com/login.aspx?direct=true&AuthType=ip,sso&db=c9h&AN=122251238&site=eds-live&scope=site

Uccheddu, S., Albertini, M., Pierantoni, L., Fantino, S., & Pirrone, F. (2019). The impacts of a reading-to-dog programme on attending and reading of nine children with autism spectrum disorders. *Animals, 9*(8), 491, 2076-2615 https://ezproxy.gvsu.edu/login?

url=https://search.ebscohost.com/login.aspx?direct=true&AuthType=ip,sso&db=asn&AN=138942296&site=eds-live&scope=site
Uglow, L. (2019). The benefits of an animal-assisted intervention service to patients and staff at a children's hospitalAnimal. *British Journal of Nursing, 28*(8), 509–515.
Vidal, R., Vidal, L., Ristol, F., Domènec, E., Segú, M., Vico, C., Gomez-Barros, N., & Ramos-Quiroga, J. A. (2020). Dog-assisted therapy for children and adolescents with fetal Alcohol spectrum disorders a randomized controlled pilot study. *Frontiers in Psychology, 11*. https://doi.org/10.3389/fpsyg.2020.01080
Villafaina-Domínguez, B., Collado-Mateo, D., Merellano-Navarro, E., & Villafaina, S. (2020). Effects of dog-based animal-assisted interventions in prison population: A systematic review. *Animals, 10*(11), 2129, 2076-2615 https://ezproxy.gvsu.edu/login?url=https://search.ebscohost.com/login.aspx?direct=true&AuthType=ip,sso&db=asn&AN=147285691&site=eds-live&scope=site
Vincent, A., Heima, M., & Farkas, K. J. (2020). Therapy dog support in pediatric dentistry: A social welfare intervention for reducing anticipatory anxiety and situational fear in children. *Child And Adolescent Social Work Journal, 37*(6), 615–629. https://doi.org/10.1007/s10560-020-00701-4
Vodka, M. Y., Aliieva, T. A., & Vashkite, I. D. (2018). Animal-assisted therapy as a method of working in children with early mental retardation. *Актуальні Проблеми Сучасної Медицини, 1*, 42–49. https://ezproxy.gvsu.edu/login?url=https://search.ebscohost.com/login.aspx?direct=true&AuthType=ip,sso&db=edsdoj&AN=edsdoj.12b35fb86faf43d6b9d6fa3e88501971&site=eds-live&scope=site
Wolan-Nieroda, Andżelina, Dudziak, Jadwiga, Drużbicki, Mariusz, Pniak, Bogumiła, & Agnieszka Guzik. (2021). Effect of dog-assisted therapy on psychomotor development of children with intellectual disability. *Children, 8*(13). https://doi.org/10.3390/children8010013, 13–13.
Wood, W., Alm, K., Benjamin, J., Thomas, L., Anderson, D., Pohl, L., & Kane, M. (2021). Optimal terminology for services in the United States that incorporate horses to benefit people: A consensus document. *Journal of Alternative and Complementary Medicine (New York, N.Y.), 27*(1), 88–95. https://doi.org/10.1089/acm.2020.0415
Zhang, Y., Yan, F., Li, S., Wang, Y., & Ma, Y. (2021). Effectiveness of animal-assisted therapy on pain in children: A systematic review and meta-analysis. *International Journal of Nursing Sciences, 8*(1), 30–37. https://doi.org/10.1016/j.ijnss.2020.12.009

CHAPTER 10

Animal-assisted therapy for cardiac conditions*

Sami V. Abate
Inspira Health, AVP Clinical Research and Nursing Excellence, Vineland, NJ, United States

Prevalence of heart disease and heart failure

According to its most recent data, the Centers for Disease Control continue to report that heart disease is the leading cause of death for both men and women in the United States. More than 650,000 Americans die annually from heart disease; this level of mortality equates to nearly one in four of all deaths or one death every 30 seconds. One of the most common types of heart disease is coronary artery disease, a condition which impacts the blood vessels that supply blood and oxygen to the muscle of the heart. More than 6% of all Americans over the age of 20 have some form of coronary artery disease (Centers for Disease Control and Prevention, 2021). Another common cardiac condition is heart failure. Recent data from the Cleveland Clinic show that over one half of a million new cases of heart failure is diagnosed annually and more than five million Americans are currently living with varying degrees of heart failure. Among people aged 65 and older, heart failure was, in data collected before the COVID-19 pandemic, the leading cause of hospital admission (Cleveland Clinic, 2022).

Heart failure pathophysiology and management

Heart failure results when the heart's ability to effectively pump blood has been impaired by any number of causes including coronary artery disease, damage caused by heart attacks (myocardial infarctions), and cardiomyopathy (enlargement and weakening of the heart muscle). Hypertension (high

* This chapter is dedicated to the memory of West Virginia Gypsy (2003–2016), a retired (not rescued) racing greyhound who, while she never won a race, was a hero to the patients whose hearts she helped heal. The compelling evidence she helped create informed many of the works cited in this chapter, further evidencing the extraordinary capacity therapy dogs have to help mend hearts.

Animal Assisted Therapy Use Application by Condition
ISBN 978-0-323-98815-5
https://doi.org/10.1016/B978-0-323-98815-5.00010-0

blood pressure), the long-term effects of diabetes, heart defects, obesity, abnormal heart rhythms (arrhythmias, such as atrial fibrillation), and even some medications (such as certain types of chemotherapy) can also contribute to heart failure. The symptoms of heart failure can range from mild exercise intolerance to debilitating shortness of breath (caused by a buildup fluid in the lungs) and edema (swelling). Often, heart failure is managed, rather than cured. The management plan for patients with heart failure is typically multifaceted and includes several medical as well as lifestyle interventions. A number of these treatment strategies such as diuretic therapy and dietary management require commitment on the part of the patient as they are not without unpleasant side effects (such as frequent urination resulting from diuretic use and a preference for freshly prepared foods over prepackaged foods to minimize sodium intake). Heart failure patients require close monitoring by their healthcare providers and are prone to exacerbations which often require both swift recognition and early, targeted intervention (Yancy et al., 2017). Other mitigation measures require patients to closely monitor their intake and output, weigh themselves regularly to assess for fluid retention, and engage in physical activity, all of which can be taxing for patients who are short of breath. Because of the fragile nature of heart failure, quality of life is a particular concern for heart failure patients. This is especially relevant because recent research has shown that Health-Related Quality of Life is "a strong and independent predictor of all-cause death and heart failure hospitalization" in virtually all heart failure patients regardless of where they live, how severe they symptoms are, and how impaired their heart function is (Johansson et al., 2021).

Animal-assisted interventions and cardiac outcomes

Why, in particular, are animal-assisted interventions so uniquely suited to patients with heart disease? There are several potential mechanisms of action which may result in this phenomenon. First, animal-assisted interventions have demonstrated, repeatedly, that they can help regulate heart rate and blood pressure. Second, there is a growing body of research which links the use of animal-assisted interventions to the mitigation of both stress and the body's hormonal response to stress. And finally, there is a not insignificant body of work which demonstrates that the presence of companion animals improves engagement in key mitigators of heart disease including physical activity, obesity, and hyperlipidemia (Levine et al., 2013).

The nature of the relationship between human wellness and the companionship of animals, most often dogs, has been the source of both intrigue and study for generations (Hooker et al., 2002; Machová et al., 2019). Beginning in the 1980s, compelling, and sometimes high-profile, work has linked the mere ownership of a dog with a decrease in several critical risk factors for cardiac disease in addition to mitigation of cardiac conditions themselves (Friedmann et al., 1980). Work in this area has continued to grow in both volume and rigor through the close of the 20th century and into the 21st century. In 2013, the American Heart Association published a scientific statement which concluded that pet ownership, especially dog ownership, is "probably associated with decreased cardiovascular disease risk" and "may have some causal role in reducing cardiovascular disease risk" (Levine et al., 2013). Based on these findings and spurred on by their own positive experiences with therapy animals, researchers are working toward establishing a robust evidence-based foundation for the application of animal-assisted interventions in various cardiovascular populations including those who already have cardiac disease and those who are at risk for cardiac disease. The literature compiled in this chapter summarizes this work and makes a compelling case for the widespread availability of animal-assisted interventions for cardiac patients as well as those at risk for cardiac disease. Further, the numerous potential mechanisms of action are compiled into an exemplar of the application of animal-assisted interventions in a specific, prevalent, and complicated cardiac disease—heart failure (American Heart Association, 2017).

Animal-assisted interventions and heart rate modulation

The association between the presence of a companion animal and heart rate has been well documented. Heart rate was among one of the first entirely objective metrics utilized to measure the impact of exposure to a therapy animal (a therapy dog). Researchers have established this link through simple measures of apical or distal pulse rate and through more complicated assessments of heart rate variability (Coakley et al., 2021). In a recent survey of patients hospitalized in general and oncologic surgical units, a 15-minute therapy dog visit resulted in statistically significant decreases in not only heart rate but also respiratory rate. These results were found in a diverse baseline population with a mean age of 55, but whose ages ranged from 18 to 80 and when baseline measurements were obtained 10 minutes before the interaction and postmeasurements were obtained immediately after the interaction (Coakley et al., 2021). Although the prevalence of heart disease in younger populations is growing in response to the increasing frequency of common risk factors (such as obesity and hyperlipidemia) occurring in these patients, cardiovascular risk increases most notably with age. For example, in the United States, the prevalence of heart disease is approximately 40% for patients aged 40 to 59 years, while it is 75% for patients aged 60 to 79 years. That prevalence increases to 86% in patients who are more than 80 years old (Rodgers et al., 2019).

Considering the increased burden of heart disease in the elderly, research examining the impact of animal-assisted interventions on the cardiac metrics of elderly patients is of particular interest. Compelling evidence has emerged from a Swedish study which showed that higher-risk elderly patients (living in a nursing home) also experienced decreases in heart rate in association with a therapy dog visit. In this study, researchers compared both heart rate and blood pressure between a cohort of patients exposed to two visits per week from a therapy dog for a period of four weeks and a control group (Handlin et al., 2018). Participants across groups fell almost entirely into the highest age risk group for cardiac disease (Rodgers et al., 2019). Specifically, participants in the animal-assisted intervention group ranged in age from 75 to 100 years old, while participants in the control group ranged in age from 78 to 85 years old. Of note, participants were further stratified by their baseline systolic blood pressures to assess outcomes in those with elevated baseline readings (>130 mmHg) and those without elevated baseline blood pressure measures (<130 mmHg). All participants in the animal-assisted intervention group spent an equal amount of time with the same therapy dog and were permitted to interact with the dog however they preferred. Participants in the control group, by contrast, did not interact with researchers outside of the context of having their data collected (i.e., there was no effort to engage in conversation outside of answering any questions posed by the participants). Among the patients who engaged in visits with the therapy dog, significant decreases were noted in heart rate between the baseline measure and the first measurement (after two weeks of therapy dog visits) and between the baseline measurement and the end of the study (after four weeks of therapy dog visits). In the subgroup of patients with higher baseline systolic blood pressure readings (>130 mmHg), significant decreases in both heart rate and blood pressure were noted (Handlin et al., 2018). These findings mimic previous findings that examined the impact of companion animals on the health outcomes of older adults. As reported in a recent systemic review of the literature, interaction with companion animals (most often dogs) resulted in increased physical activity, improved blood pressure and heart rate, as well numerous positive behavioral health outcomes (Hughes et al., 2020). While, as previously explained, the bulk of the burden of cardiac disease remains on the elderly, a growing body of work has shown a positive link between animal-assisted interventions and cardiac measures of younger adults as well. For example, in a convenience sample of college nursing students, researchers found that across the course a class session, students' average heart rates increased to a peak around the midpoint of the class then decreased nearly back down to preclass levels. When the peak and trough heart rate measurements were averaged to establish one mean heart rate per student for the class, notably (although not statistically significant) lower heart rates were seen in the students whose classroom had a therapy dog present (Griscti & Camilleri, 2020).

Animal-assisted interventions and stress mitigation

The negative health effects of stress are both well known and well documented. In adult patients, for example, major life stressors often precede subsequent diagnoses of major depression, anxiety, or even posttraumatic stress disorder. Physiologic and biologic responses to stress impact numerous body systems including the endocrine system, nervous system, immune system, and the cardiovascular system. The endocrine system, for example, responds to acute stress with the production of both epinephrine and cortisol, both of which can have cardiac consequences. The cumulative impact of chronic stress can not only suppress immune system function but also result in increased blood pressure and subsequent vascular hypertrophy which, in turn, can lead to increased cardiac workload, ventricular hypertrophy, and atherosclerosis (Schneiderman et al., 2005).

The presence of companion animals, in particular therapy pets or therapy dogs, has long been subjectively linked with the lowering of perceived stress and the promotion of general well-being (Hooker et al., 2002; Polheber & Matchock, 2014). However, scientific evidence in this field has only begun to emerge (Giaquinto & Valentini, 2009; Hooker et al., 2002). For example, Polheber and Matchock (2014) examined the stress mitigation offered by the presence of a dog as compared to the mitigation offered by a human friend by comparing various objective and subjective measures of stress including heart rate, cortisol levels, and patient-reported outcome, which provided a subjective measure of stress. In this study, nearly 300 undergraduate students were recruited from courses in both psychiatry and statistics. All students were screened for their attitudes toward dogs before taking part; only those who were primarily favorable to dogs and did not report a phobia or dislike of dogs were included. Notably, no participants were dog owners. After the completion of baseline measures, participants were randomized to three conditions: a control group (who sat quietly in the lab for 40 minutes), the friend group (spent the 40 minutes talking with a friend of their choosing in the lab), and the dog group (who petted, talked to, and gave treats to a therapy dog in the lab for 40 minutes). After their time in the lab, participants were exposed to the Trier Social Stress Test (a simulation of a stressful social situation designed to induce a stress response). Postintervention measurements indicated that those individuals who had spent time with the therapy dog demonstrated significantly lower post-Trier Social Stress Test cortisol levels than those in the control and in the friend group. No significant difference was detected between the control and the friend group. Participants in the therapy dog group were also found to have significantly lower heart rates during the Trier Social Stress Test than the other groups (Polheber & Matchock, 2014). Improvements in heart rate and stress have also been associated with exposure to animal-assisted interventions in other student populations including nursing students (Griscti & Camilleri, 2020).

A more recent metaanalysis further supports Polheber and Matchock's (2014) findings. Across 28 papers and more than 1300 participants, animal-assisted interventions were shown to consistently moderate both subjective and objective measures of stress. In addition to an impact on heart rate, animal-assisted interventions were associated with modest (not significant) improvements in blood pressure as well. Subjectively, the introduction of an animal-assisted intervention was shown to significantly moderate self-reported anxiety and self-reported stress, particularly in adults. From an operational perspective, it was notable that these beneficial effects were at their most prominent when the participants interacted with the therapy animal individually, rather than in a group setting (Ein et al., 2018).

In 2007, the American Psychological Association (APA) named stress a major health concern in the United States. More recently (in 2020), the APA stated that "We [Americans] are facing a national mental health crisis that could yield serious health and social consequences for years to come" (American Psychological Association, 2020). According to the APA's findings, young people (aged 13 to 23) are most at risk for sustained levels of chronic stress related to the global coronavirus pandemic (American Psychological Association, 2020). Of note, research exists linking both stress and heart rate modulation with animal-assisted interventions focused specifically on participants in this age range (Griscti & Camilleri, 2020; Polheber & Matchock, 2014). However, other common facets of modern American life are also chronic inducers of stressors, regardless of generation, including healthcare, violence, climate change, and uncertainty. Socially economically challenged individuals are also at an increased risk for chronic stress (American Psychological Association, 2020). Unsurprisingly, these individuals are also at a substantially higher risk of heart disease including heart failure and chronic cardiac conditions necessitating frequent hospital admission (Schultz et al., 2018). While the cost of pet ownership can be a barrier for these particularly high-risk individuals, most animal-assisted interventions are provided by volunteer therapy animal and handler teams who are accredited and insured by nationally recognized therapy animal registering programs. In fact, one of the largest animal-assisted intervention groups in the United States, Pet Partners (formerly the Delta Society), has prioritized diversity, equity, and inclusion in their programs (Pet Partners, 2022).

Animal-assisted interventions and physical activity

Regular physical activity is considered a primary prevention measure employed in the fight against cardiovascular disease. In patients already diagnosed with cardiac disease, regular physical activity is a valuable and essential component of their treatment plan. Physical activity is also both a predictor of event-free survival and directly associated with the mitigation of disease progression for patients living with cardiac disease (Winzer et al., 2018). Extensive work has linked even minimal physical activity with substantial preventative benefit for patients at risk or diagnosed with cardiac disease. For example, small amounts of regular light walking have been shown to demonstrably reduce cardiovascular risk across diverse adult populations (Carnethon, 2009). The act of physically walking a dog has often been speculated to be one of the principal mechanisms by which pet ownership can be linked to improved cardiovascular outcomes. Multiple studies have demonstrated that dog owners of all ages, including the elderly, routinely engage in light to moderate physical activity with more regularity than nondog owners. For example, dog owners across all age groups in the United States and Australia routinely walked more frequently and for longer periods of time/longer distances than their nondog owning counterparts

(While, 2017). Among elderly patients who survived a myocardial infarction, those who walked their own dog for 15 minutes three times per day continued to demonstrate better performance on bicycle exercise tests for as long as one year after their myocardial infarction when compared with nondog owners(Ruzic et al., 2011). To capitalize on the benefits of dog walking, researchers at Purdue University recently demonstrated that dog walking can be further maximized with the addition of a motivational program that provided dog owners with routine email reminders. Dog owners who participated in the motivational intervention walked significantly longer (almost one full hour more) each week than those dog owners who did not receive the email motivators (Richards et al., 2016).

Could this dog-associated promotion of physical activity be extended beyond dog owners via animal-assisted therapy interventions? The answer appears to be a resounding yes! For example, when an animal-assisted intervention component was added to a poststroke gait training program, participants (a cohort of Korean patients whose mean age was in their sixties) in the interventional arm demonstrated more improvement than those who participated in the traditional program. More specifically, those participants whose gait training was augmented by the presence of a therapy dog showed greater improvement in the areas of cadence, speed, length of stride, and gait symmetry (An & Park, 2021). Young people have also demonstrated a response to the addition of a therapy dog into programs designed to promote their physical activity. This activity promoting phenomenon has been shown in both well children and in adolescents with medical conditions that impeded normal mobility. For example, physical activity levels increased in a small cohort ($n = 7$) of adolescents with slipped capitol femoral epiphyses or Blount's disease (a condition which causes gait abnormalities and can cause pain with ambulation). Not only did the participants unanimously report enjoying their experience walking with a therapy dog but also they reported engaging more physical activity during the therapy walking sessions as well as after (Vitztum et al., 2016). Similarly, in a small group of obese children ($n = 12$) aged eight to 12, the presence of a therapy dog was shown to be a motivator of physical activity when compared to a human therapy leader alone. In the study condition, the children spent more time completing agility exercises when they had a chance to attempt to outrun the dog (vs. a human facilitator) and completed more passes of a retrieving game (in which they raced to a base while either the human facilitator or the dog retrieved a thrown object) when playing with the dog (Wohlfarth et al., 2013).

Exemplar: the application of animal-assisted interventions in heart failure

Can a therapy dog help heal a broken heart? Two studies, conducted on opposite sides of the United States appear to indicate that they can. The two studies described in detail here indicate that even brief interactions with therapy dogs can markedly impact the outcomes of both the most acutely ill heart failure patients as well as those recovering from an exacerbation of a chronic condition.

Researchers at the University of Los Angeles recognized both the prevalence and complexity of severe heart failure. In particular, the researchers recognized the deleterious effects that the chronic strain heart failure can cause, including activation of the neuroendocrine cascade and sympathetic nervous system which can result in ventricular remodeling, thus worsening cardiac function. In recognition of the stress-mitigating effects of animal-assisted interventions and the positive impact exposure to companion animals can have on cardiac parameters (including heart rate and blood pressure), the researcher sought to study the benefits of a brief (12 minute) therapy dog intervention on critically ill heart failure patients. A sample of 76 patients whose heart failure required intensive medical management which necessitated the placement of a pulmonary artery catheter (a catheter whose tip lies in the pulmonary artery and that can report pressure and other hemodynamic metrics inside of the heart) was randomized to three groups: a control group (who received the standard of care), a volunteer group (who received a brief visit from a volunteer alone), and an intervention group (who received a 12-minute visit from a therapy dog team). Measures were conducted at baseline, after eight minutes of the intervention (or control condition) and at 16 minutes (four minutes after conclusion of the intervention). The comprehensive cadre of measures included heart rate, blood pressure, cardiac index (a measure of blood flow calculated by dividing cardiac output by body surface area), pulmonary artery pressures, pulmonary capillary wedge pressure, catecholamine levels (epinephrine and norepinephrine, drawn from the pulmonary artery catheter, not from a venipuncture), and anxiety (measured using the Spielberger State Trait Anxiety Index). When compared to the volunteer group, participants in the animal-assisted intervention group demonstrated significantly greater decreases in systolic pulmonary artery pressure, pulmonary capillary wedge pressure, epinephrine levels, and norepinephrine levels. When compared to the control group, participants in the animal-assisted intervention group had significantly greater decreases in systolic pulmonary artery pressure and pulmonary capillary wedge pressure. In addition to these physiologic parameters, participants in the animal-assisted intervention group also demonstrated the most marked decreases in state anxiety, with significantly greater decreases in anxiety seen in the animal-assisted intervention group when compared to the volunteer group (Cole et al., 2007). This study was, at the time of publication, the first study to potentially document the source of the cardiac benefits often observed in pet owners and in the recipients of animal-assisted interventions by literally measuring the impact of exposure to a companion animal on the inner workings of acutely failing hearts.

As previously described, heart failure is a prevalent and complex condition that necessitates a multifaceted management strategy which requires the active engagement of the patient. To that end, one regional medical center in rural southern New Jersey implemented a multidisciplinary heart failure management program designed to teach hospitalized heart failure patients about their conditions and engage them, during their hospital stay, in the self-care interventions that would help improve their outcomes after discharge. One component of the program employed specially trained restorative aides whose primary function was to engage patients in early ambulation. While the program was well received and successful, ambulation remained a pain point with many patients refusing the opportunity/request to ambulate despite a personalized request to do so with individualized support. This was especially concerning given the established link between early ambulation and future cardiac outcomes in heart failure patients (Fleming et al., 2018; Jolly et al., 2007). To help overcome these challenges, Abate et al. (2011) introduced a novel canine-assisted ambulation (CAA) program to help promote early ambulation among heart failure patients. Sixty-nine admitted patients were recruited for the pilot study. The outcomes of these intervention patients were ultimately compared with the outcomes of a matched sample 69 heart failure patients from the organizational database of patients who had received the standard of care (including interaction with the restorative aides). To minimize bias, patients being seen by the restorative aides were asked if they would like to ambulate and their response was recorded according to standard practice (with no mention of the therapy dog). Patients were then assessed for fear or dislike of dogs by the restorative aide and any potential allergies to dogs were also assessed. If no allergies, fear, or dislike were noted, the patients were then asked if they would like the chance to meet and walk with a therapy dog (which was still out of sight). Those who agreed to walking with the dog were given the chance to walk with the dog and the restorative aide (as well as the therapy dog handler) for as long as they would like. The distance ambulated was recorded in steps. After the ambulation was completed and the patient had an opportunity to further interact with the therapy dog if desired, the study was explained, and patients were asked for their consent to have their outcomes included in the study. Consenting patients had their ambulation outcomes entered into the study data base and completed a brief instrument to assess their satisfaction with the intervention. Analysis of the results showed that while 38 of 69 patients initially refused ambulation, a significant number of these 38 patients (n = 13) agreed to walk when offered the chance to do so with the therapy dog. Patients who walked with the therapy dog walked significantly farther than similar patients in the comparison group, and all patients responded favorably to the opportunity to walk with a therapy dog. Notably, the lowest scoring question on the satisfaction survey was the question which assessed patients' assessment of whether they walked further because of the dog indicating that many patients may not have realized how far they walked because their attention was focused on the therapy dog. Patients in the CAA group were also discharged from the hospital one day sooner than the patients in the matched comparison group (Abate et al., 2011).

In both studies reported here, none of the heart failure patients experienced any adverse events despite the physical nature of the interaction between the patient and the therapy dog. In the case of the severe heart failure study, the therapy dog was permitted to lie in the bed with the patient. In the case of the CAA study, the patients walked alongside of the therapy dog (Abate et al., 2011; Cole et al., 2007). In combination, these two studies provide a compelling case for the utilization of animal-assisted interventions across the breadth and scope of heart failure care. Heart failure, like many cardiac diseases, is often characterized by a series of exacerbations and remissions and by the need for lifelong chronic disease management. These studies document the acute and real-time benefits exposure to a therapy dog can have and the potential long-term recovery benefits that the addition of a therapy dog to the standard of care can offer. Because the role of self-management is so integral in many cardiac conditions, patient education programs may be another area of opportunity for animal-assisted interventions in the future. Existing work has already indicated that, in addition to mitigating stress, the availability of a therapy dog may promote a relaxing atmosphere that facilitates attention in adults learning healthcare-focused material (Griscti & Camilleri, 2020). Examining all of these results in combination, and while considering the impact of animal-assisted interventions on heart rate, stress levels, and physical activity and with the knowledge of the long-term survival benefits that can be associated with exposure to companion animals, the case for therapy dogs having the ability to literally mend broken hearts is persuasive.

Animal-assisted interventions and cardiac outcomes in unique populations

The animal-assisted intervention mechanisms of action described here (positive impacts on subjective and objective markers of stress, heart rate, blood pressure, and physical activity) are also applicable to other, more unique, patient populations impacted by cardiovascular diseases beyond heart failure. This is particularly true in pediatric patients. While cardiovascular disease, as previously described, is primarily a condition that disproportionally impacts the elderly, pediatric cardiac disease can be both devastating and debilitating. Globally, more than 200,000 pediatric deaths were attributed in 2017 to congenital heart disease; nearly 70% of these deaths occurred in infants less than one year of age. While the survival of congenital heart disease is increasing, longer survival also equates to an increased number of children and adolescents living with the burden of cardiac disease (Zimmerman et al., 2020).

Researchers at a children's hospital in Arkansas (United States) have recently explored the addition a therapy dog to postoperative therapy routines of adolescent patients who had received a heart transplant. While participants in the intervention group did not walk as far as those in the control group (approximately 27 feet less), they did spend, on average, an additional two minutes walking when the therapy dog was present. Perhaps most important to this feasibility study was the fact that the participants enjoyed the time with the therapy dog, all participant's vital signs remained stable, and that no adverse or safety events were reported (Walden et al., 2020). Other pediatric studies have had equally promising results which may be translated into adult cardiac populations and other cardiac diseases. For example, researchers at a pediatric hospital in Canada found that a visiting therapy animal resulted in a number of positive outcomes on a pediatric inpatient cardiology unit. The presence of the hospital's therapy dog on the unit was shown to reduce stress among patients and their parents, thus improving the therapeutic milieu of the unit. Changes in respiratory and heart rate over the course of the visits also indicated that the therapy dog visits promoted relaxation. Notably, all participants in the study requested additional therapy dog visits and the degree of rapport and interaction with the therapy dog was positively correlated to the benefits the participant experienced (Wu et al., 2002). Physiologic measures associated with cardiac disease have also been observed to improve in pediatric postoperative patients (who had surgery for noncardiac conditions). The introduction of a therapy dog at hour two of a child's postoperative period resulted in lower blood pressures, higher oxygenation saturation levels, and higher cerebral oxygenation measurements in the animal-assisted intervention group when compared to those patients who received the standard of care. In addition, after the introduction of the therapy dog, all of the patients demonstrated diffuse beta activity (a marker of alertness) on encephalography whereas none was observed in the control group. Again, despite the vulnerable nature of these patients, no adverse events were reported (Calcaterra et al., 2015). The possibility exists that perhaps these findings may be replicated in a larger and more diverse populations of cardiac and noncardiac surgical patients and in both adult and pediatric patients. Further, when coupled with the known mitigating impacts that animal-assisted interventions can have on both acute (including postoperative) and chronic pain, these findings make for a promising line of future cardiology-focused study (Harper et al., 2015; Rodrigo-Claverol et al., 2019).

Looking back and forward: the heart of animal-assisted interventions

The impetus for the CAA study described in detail here was an interaction between the same therapy dog (a retired racing greyhound named Gypsy, formerly West Virginia Gypsy) and a minimally responsive patient with severe developmental disabilities who was at the end of life. The subtle, but noticeable, signs of engagement observed by the therapy dog volunteer team prompted the team to work with organizational leaders to explore other potential opportunities to both offer and study animal-assisted interventions in the organization. More than a decade later (and more than 8000 miles away), Swiss researchers Hedigre et al. (2019) documented what the New Jersey team had witnessed. The Swiss team recorded an increase in random eye movements and other indicators of higher levels of consciousness in minimally conscious patients during animal-assisted interventions than what could be documented at baseline. While controversial and not generally accepted as meaningful, indicators of greater arousal and alertness were also seen in analysis of these subjects' high-frequency and low-frequency heart rate variability parameters (Hediger et al., 2019).

The results of the various cardiology-focused and cardiology-adjacent studies reviewed in this chapter demonstrate the exceptional promise animal-assisted interventions can offer to cardiology patients across all phases of the cardiac disease continuum and across the entire life span. For millennia, leading thinkers including Aristotle himself viewed the heart as the center of emotion. Popular and secular culture has been slower to accept the brain as the seat of cognition and emotion. Even among medical professionals, cessation of heart function is sometimes more immediately recognized as an indicator of death versus cessation of cognitive function (Brandt & Huppert, 2021). As demonstrated in this chapter, there is a nearly inextricable link between the emotion well-being of cardiac patients and the physical wellness of cardiac patients. Both goal-directed animal-assisted interventions and the simple act of interacting with a therapy animal offer a promising (and well-documented) bridge between the two tangled elements. Essentially, a compelling body of work now exists that almost indisputably shows that animal-assisted interventions can safely and effectively facilitate keeping a heart healthy and aid in mending a literally broken heart. As healthcare providers, therapy animal volunteers, and students of the science of animal-assisted interventions, we are obligated to explore all possible avenues to improve the lives of those we serve. In the pursuit of evidence-based best practices, future work that examines the impact of animal-assisted interventions on cardiac conditions and the conditions and circumstances that mitigate them are very much needed. This work should be conducted with diverse patient populations, in varied clinical and nontraditional healthcare settings, and address populations negatively impacted by disparities in healthcare access. The remote monitoring and telehealth modalities made ubiquitous during the pandemic and the contemporaneous proliferation of wearable technology offer new options for data collection that may help researchers to better document the physiologic impacts of therapy animal interventions. Finally, as additional research in the cardiac field (and in all fields) strengthens the evidence base in support of the safety and efficacy of animal-assisted interventions, the barriers which preclude the adoption of animal-assisted interventions should simultaneously be lessened.

This author once referred to the therapy dog Gypsy as a "heart-healing hound." In doing so, she was referring to the comfort Gypsy's presence brought in a time of "heartbreaking" personal loss. The evidence presented here suggests that therapy dogs around the world have the potential to physically and actually heal broken hearts. Considering the breadth of burden cardiac disease places on the world's population, their services are greatly needed.

References

Abate, S. V., Zucconi, M., & Boxer, B. A. (2011). Impact of canine-assisted ambulation on hospitalized chronic heart failure patients' ambulation outcomes and satisfaction. *Journal of Cardiovascular Nursing, 26*(3), 224–230. https://doi.org/10.1097/jcn.0b013e3182010bd6

American Heart Association. (2017). What is heart failure. www.heart.org/en/health-topics/heart-failure/what-is-heart-failure.

American Psychological Association. (2020). *Stress in America: A national mental health crisis.* www.apa.org/news/press/releases/stress/2020/report-october.

An, H. J., & Park, S. J. (2021). Effects of animal-assisted therapy on gait performance, respiratory function, and psychological variables in patients post-stroke. *International Journal of Environmental Research and Public Health, 18*(11). https://doi.org/10.3390/ijerph18115818

Brandt, T., & Huppert, D. (2021). Brain beats heart: A cross-cultural reflection. *Brain, 144*(6), 1617–1620. https://doi.org/10.1093/brain/awab080

Calcaterra, V., Veggiotti, P., Palestrini, C., De Giorgis, V., Raschetti, R., Tumminelli, M., Mencherini, S., Papotti, F., Klersy, C., Albertini, R., Ostuni, S., Pelizzo, G., & Schwentner, C. (2015). Post-operative benefits of animal-assisted therapy in pediatric surgery: A randomised study. *PLOS one, 10*(6), e0125813. https://doi.org/10.1371/journal.pone.0125813

Carnethon, M. R. (2009). Physical activity and cardiovascular disease: How much is enough? *American Journal of Lifestyle Medicine, 3*(1_Suppl. l), 44S–49S. https://doi.org/10.1177/1559827609332737

Centers for Disease Control and Prevention. (2021). *Heart disease facts.* www.cdc.gov/heartdisease/facts.htm.

Cleveland Clinic. (2022). *Heart failure.* https://my.clevelandclinic.org/health/diseases/17069-heart-failure-understanding-heart-failure.

Coakley, A. B., Annese, C. D., Empoliti, J. H., & Flanagan, J. M. (2021). The experience of animal assisted therapy on patients in an acute care setting. *Clinical Nursing Research, 30*(4), 401–405. https://doi.org/10.1177/1054773820977198

Cole, K. M., Gawlinski, A., Kotlerman, J., & Steers, N. (2007). Animal-assisted therapy in patients hospitalized with heart failure. *American Journal of Critical Care, 16*(6), 575–585. http://ajcc.aacnjournals.org/cgi/content/full/16/6/575.

Ein, N., Li, L., & Vickers, K. (2018). The effect of pet therapy on the physiological and subjective stress response: A meta-analysis. *Stress and Health, 34*(4), 477–489. https://doi.org/10.1002/smi.2812

Fleming, L. M., Zhao, X., DeVore, A. D., Heidenreich, P. A., Yancy, C. W., Fonarow, G. C., Hernandez, A. F., & Kociol, R. D. (2018). Early ambulation among hospitalized heart failure patients is associated with reduced length of stay and 30-day readmissions. *Circulation: Heart Failure, 11*(4). https://doi.org/10.1161/CIRCHEARTFAILURE.117.004634

Friedmann, E., Katcher, A. H., Lynch, J. J., & Thomas, S. A. (1980). Animal companions and one-year survival of patients after discharge from a coronary care unit. *Public Health Reports, 95*(4), 307–312.

Giaquinto, S., & Valentini, F. (2009). Is there a scientific basis for pet therapy. *Disability and Rehabilitation, 31*(7), 595–598. https://doi.org/10.1080/09638280802190735

Griscti, O., & Camilleri, L. (2020). The impact of dog therapy on nursing students' heart rates and ability to pay attention in class. *International Journal of Educational Research, 99*, 101498. https://doi.org/10.1016/j.ijer.2019.101498

Handlin, L., Nilsson, A., Lidfors, L., Petersson, M., & Uvnäs-Moberg, K. (2018). The effects of a therapy dog on the blood pressure and heart rate of older residents in a nursing home. *Anthrozoös, 31*(5), 567–576. https://doi.org/10.1080/08927936.2018.1505268

Harper, C. M., Dong, Y., Thornhill, T. S., Wright, J., Ready, J., Brick, G. W., & Dyer, G. (2015). Can therapy dogs improve pain and satisfaction after total joint arthroplasty? A randomized controlled trial. *Clinical Orthopaedics and Related Research, 473*(1), 372–379. https://doi.org/10.1007/s11999-014-3931-0

Hediger, K., Petignat, M., Marti, R., Hund-Georgiadis, M., & Martinuzzi, A. (2019). Animal-assisted therapy for patients in a minimally conscious state: A randomized two treatment multi-period crossover trial. *PLOS One, 14*(10), e0222846. https://doi.org/10.1371/journal.pone.0222846

Hooker, S. D., Holbrook Freeman, L., & Stewart, P. (2002). Pet therapy research. *Holistic Nursing Practice, 17*(1), 17–23. https://doi.org/10.1097/00004650-200210000-00006

Hughes, M. J., Verreynne, M. L., Harpur, P., & Pachana, N. A. (2020). Companion animals and health in older populations: A systematic review. *Clinical Gerontologist, 43*(4), 365–377. https://doi.org/10.1080/07317115.2019.1650863

Johansson, I., Joseph, P., Balasubramanian, K., McMurray, J. J. V., Lund, L. H., Ezekowitz, J. A., Kamath, D., Alhabib, K., Bayes-Genis, A., Budaj, A., Dans, A. L. L., Dzudie, A., Probstfield, J. L., Fox, K. A. A., Karaye, K. M., Makubi, A., Fukakusa, B., Teo, K., Temizhan, A., … Yusuf, S. (2021). Health-related quality of life and mortality in heart failure: The global congestive heart failure study of 23000 patients from 40 countries. *Circulation*, 2129–2142. https://doi.org/10.1161/CIRCULATIONAHA.120.050850

Jolly, K., Taylor, R., Lip, G. Y. H., Greenfield, S., Raftery, J., Mant, J., Lane, D., Jones, M., Lee, K. W., & Stevens, A. (2007). The Birmingham Rehabilitation Uptake Maximisation Study (BRUM). Home-based compared with hospital-based cardiac rehabilitation in a multi-ethnic population: Cost-effectiveness and patient adherence. *Health Technology Assessment, 11*(35). https://doi.org/10.3310/hta11350

Levine, G. N., Allen, K., Braun, L. T., Christian, H. E., Friedmann, E., Taubert, K. A., Thomas, S. A., Wells, D. L., & Lange, R. A. (2013). Pet ownership and cardiovascular risk: A scientific statement from the American Heart Association. *Circulation, 127*(23), 2353–2363. https://doi.org/10.1161/CIR.0b013e31829201e1

Machová, K., Procházková, R., Eretová, P., Svobodová, I., & Kotík, I. (2019). Effect of animal-assisted therapy on patients in the department of long-term care: A pilot study. *International Journal of Environmental Research and Public Health, 16*(8), 1362. https://doi.org/10.3390/ijerph16081362

Pet Partners. (2022). *Pet partners: About us*. https://petpartners.org/about-us/.

Polheber, J. P., & Matchock, R. L. (2014). The presence of a dog attenuates cortisol and heart rate in the Trier Social Stress Test compared to human friends. *Journal of Behavioral Medicine, 37*(5), 860–867. https://doi.org/10.1007/s10865-013-9546-1

Richards, E. A., Ogata, N., & Cheng, C.-W. (2016). Evaluation of the dogs, physical activity, and walking (dogs PAW) intervention. *Nursing Research, 65*(3), 191–201. https://doi.org/10.1097/nnr.0000000000000155

Rodgers, J. L., Jones, J., Bolleddu, S. I., Vanthenapalli, S., Rodgers, L. E., Shah, K., Karia, K., & Panguluri, S. K. (2019). Cardiovascular risks associated with gender and aging. *Journal of Cardiovascular Development and Disease, 6*(2). https://doi.org/10.3390/jcdd6020019

Rodrigo-Claverol, M., Casanova-Gonzalvo, C., Malla-Clua, B., Rodrigo-Claverol, E., Jové-Naval, J., & Ortega-Bravo, M. (2019). Animal-assisted intervention improves pain perception in polymedicated geriatric patients with chronic joint pain: A clinical trial. *International Journal of Environmental Research and Public Health, 16*(16), 2843. https://doi.org/10.3390/ijerph16162843

Ruzic, A., Milteic, B., Ruzic, T., Persic, V., & Laskarin, G. (2011). Regular dog-walking improves physical capacity in elderly patients after myocardial infarction. *Collegium Antropologicum, 35*(Suppl. l_2), 73–75.

Schneiderman, N., Ironson, G., & Siegel, S. D. (2005). Stress and health: Psychological, behavioral, and biological determinants. *Annual Review of Clinical Psychology, 1*, 607–628. https://doi.org/10.1146/annurev.clinpsy.1.102803.144141

Schultz, W. M., Kelli, H. M., Lisko, J. C., Varghese, T., Shen, J., Sandesara, P., Quyyumi, A. A., Taylor, H. A., Gulati, M., Harold, J. G., Mieres, J. H., Ferdinand, K. C., Mensah, G. A., & Sperling, L. S. (2018). Socioeconomic status and cardiovascular outcomes: Challenges and interventions. *Circulation, 137*(20), 2166–2178. https://doi.org/10.1161/CIRCULATIONAHA.117.029652

Vitztum, C., Kelly, P. J., & Cheng, A. L. (2016). Hospital-based therapy dog walking for adolescents with orthopedic limitations: A pilot study. *Comprehensive Child and Adolescent Nursing, 39*(4), 256–271. https://doi.org/10.1080/24694193.2016.1196266

Walden, M., Lovenstein, A., Randag, A., Pye, S., Shannon, B., Pipkin, E., Ramick, A., Helmick, K., & Strickland, M. (2020). Methodological challenges encountered in a study of the impact of animal-assisted intervention in pediatric heart transplant patients. *Journal of Pediatric Nursing, 53*, 67–73. https://doi.org/10.1016/j.pedn.2020.04.017

While, A. (2017). Pet dogs as promoters of wellbeing. *British Journal of Community Nursing, 22*(7), 332–336. https://doi.org/10.12968/bjcn.2017.22.7.332

Winzer, E. B., Woitek, F., & Linke, A. (2018). Physical activity in the prevention and treatment of coronary artery disease. *Journal of the American Heart Association, 7*(4). https://doi.org/10.1161/JAHA.117.007725

Wohlfarth, R., Mutschler, B., Beetz, A., Kreuser, F., & Korsten-Reck, U. (2013). Dogs motivate obese children for physical activity: Key elements of a motivational theory of animal-assisted interventions. *Frontiers in Psychology, 4*. https://doi.org/10.3389/fpsyg.2013.00796

Wu, A. S., Niedra, R., Pendergast, L., & McCrindle, B. W. (2002). Acceptability and impact of pet visitation on a pediatric cardiology inpatient unit. *Journal of Pediatric Nursing, 17*(5), 354–362. https://doi.org/10.1053/jpdn.2002.127173

Yancy, C. W., Jessup, M., Bozkurt, B., Butler, J., Casey, D. E., Colvin, M. M., Drazner, M. H., Filippatos, G. S., Fonarow, G. C., Givertz, M. M., Hollenberg, S. M., Lindenfeld, J. A., Masoudi, F. A., McBride, P. E., Peterson, P. N., Stevenson, L. W., & Westlake, C. (2017). 2017 ACC/AHA/HFSA focused update of the 2013 ACCF/AHA guideline for the management of heart failure: A report of the American College of Cardiology/American Heart Association task force on Clinical Practice guidelines and the heart failure society of America. *Circulation, 136*(6), e137–e161. https://doi.org/10.1161/CIR.0000000000000509

Zimmerman, M. S., Smith, A. G. C., Sable, C. A., Echko, M. M., Wilner, L. B., Olsen, H. E., Atalay, H. T., Awasthi, A., Bhutta, Z. A., Boucher, J. L., Castro, F.,

Cortesi, P. A., Dubey, M., Fischer, F., Hamidi, S., Hay, S. I., Hoang, C. L., Hugo-Hamman, C., Jenkins, K. J., … Kassebaum, N. J. (2020). Global, regional, and national burden of congenital heart disease, 1990–2017: A systematic analysis for the global burden of disease study 2017. *The Lancet Child & Adolescent Health, 4*(3), 185–200. https://doi.org/10.1016/s2352-4642(19)30402-x

CHAPTER 11

Animal-assisted therapy in geriatric patients

Dominic M. DeMarco[1], Kala Chinnaswamy[1] and George T. Grossberg[2]
[1]Saint Louis University School of Medicine, Department of Psychiatry and Behavioral Neuroscience, Saint Louis, MO, United States; [2]Saint Louis University School of Medicine, Division of Geriatric Psychiatry, Department of Psychiatry and Behavioral Neuroscience, Saint Louis, MO, United States

Animal-assisted therapy (AAT) has multiple indications in the older adult population in terms of both physical and mental health (Cherniack & Cherniack, 2014). There may be a biochemical basis for this, as ongoing contact with pets has been shown to increase concentrations of β-endorphin, oxytocin, prolactin, β-phenylethylamine, and dopamine, as well as to reduce plasma levels of cortisol (Odendaal & Meintjes, 2003). AAT has also been shown to have other quality-of-life benefits, such as improving gait, balance, and communication among older adults interacting with animals (Rodrigo-Claverol et al., 2020).

Much of the utility of AAT in older adults has been observed in patients with psychiatric disorders, particularly in patients with dementia. Management of dementia patients presents a unique challenge as they may exhibit difficulties in multiple cognitive domains including social interaction and communication, mood, and behavior. Additionally, loneliness and depression are prevalent issues in this population, and these factors themselves have been established as independent risk factors for dementia (Byers & Yaffe, 2011; Wilson et al., 2007). Management of older adults with dementia is often multifaceted, and it involves both nonpharmacologic and pharmacologic interventions. Of most concern is behavioral management, which can often involve the use of antipsychotic medications (Desai et al., 2012). Given the potential side effects and drug-interactions that come with antipsychotic medications, nonpharmacologic interventions are preferred as the initial approach (Jennings et al., 2018). AAT is a viable means of nonpharmacologic therapy for behavior concerns in patients with dementia, such as agitation and aggression (Majić et al., 2013).

Animal Assisted Therapy Use Application by Condition
ISBN 978-0-323-98815-5
https://doi.org/10.1016/B978-0-323-98815-5.00006-9

An analysis of existing research on AAT in older adults with psychiatric disorders was conducted by Peluso et al. in 2018 (Peluso et al., 2018). Although this review found the amount of data and conclusiveness of existing research sparse, it did suggest that AAT may have benefit in emotional, behavioral, and psychological symptoms along with improving daily living activities in these patients. Specifically, AAT seemed to promote social and communication skills, facilitate verbal and body language and interaction, increase well-being, self-esteem, and positive mental attitude, and increase the desire to participate in leisure activities and be in a group. It also suggested improved quality of life, mood (in particular, anxiety and loneliness), motivation, and social behavior with AAT.

AAT may possibly have positive effects on other psychiatric illnesses in older adults as well. For example, a recent study by Chen et al. describes a small randomized controlled trial that focused specifically on the older adult population with schizophrenia (Chen et al., 2021). This study directly compared a control group against a group of patients who had the addition of AAT to their normal treatment regimen. The main benefit of the study was the reduction of negative symptoms in the AAT group.

Depression and anxiety are also prominent issues in the older adult population. A small study by Moretti et al. (Moretti et al., 2011) found that interacting with a dog over a 90-minute period once weekly led to a decrease in the Geriatric Depression Scale in a group of institutionalized elderly patients with mental illness. Research has also identified benefits in controlling anxiety in specific circumstances that could be particularly beneficial to the older adult population. AAT has been studied and has shown benefit in hospitalized patients, large segments of whom are older adults (Waite et al., 2018). A pilot study by Smith et al. showed that a one-time, 12—20-minute session with a therapy dog reduced anxiety in geriatric patients that were hospitalized (Smith et al., 2020). This study is not only significant in that it studied an older population than previous research, having an average study population age of 80 years old, but also in that it provided preliminary evidence for a nonpharmacologic means of reducing anxiety in the inpatient setting. Research has also been done with specific hospital procedures, including electroconvulsive therapy (ECT), a method often employed to treat depression in older adults. The inherent nature of ECT can be daunting and anxiety provoking for patients, and there is evidence that AAT reduces preprocedure anxiety specifically in older adults undergoing this procedure (Barker et al., 2003).

As technology becomes increasingly part of daily life, it has also become more and more relevant to medical treatment. New research is exploring ways to bring the benefits of AAT to older adults without the additional complications of managing the animals themselves. Robotic pet companions have become an emerging solution to this problem. Like their living counterparts, robotic companions have been shown to benefit older patients, particularly in the area of elderly care and dementia (Góngora Alonso et al., 2019). These companions have been studied and implemented in a range of forms, from the classical robot PaPeRo to the animatronic seal PARO. PARO has been specifically shown in research to increase overall quality of life in older adults with dementia, as measured by QOL-AD and OERS Pleasure Scores (Moyle et al., 2013). In one small study, a robot in the form of a cat was shown to decrease agitation among older adult participants in a nursing home setting (Marsilio et al., 2018). In another survey of patient families and staff in an intensive care unit regarding a cat-modeled robotic pet given to delirious patients, 65% of subjects, their families, and clinical staff agreed that the cat was calming. Over 70% of respondents did not feel that the cat interfered with clinical care (Schulman-Marcus et al., 2019). While the study design did not test whether the robotic pet decreased pharmacotherapy use, its effects on calming previously-agitated patients show promise for additional study. These results suggest that robotic pet companions may have benefits in reducing the use of pharmacologic modalities to treat delirium. While still minimally studied at this point, preliminary findings and the fact that these robotic companions are already being used widely show that future directions toward robotic pets may be a norm for geriatric care.

While the current landscape of research into AAT continues to grow, clinical experience with AAT can provide valuable insights into the role animals can play in patient care. The following are clinical vignettes of real-world patients who have benefitted from AAT:

An 83-year-old woman was hospitalized for severe depression with catatonic features. She responded to pet therapy by cuddling a therapy kitten and uttering her first words in days

An 88-year-old man with advanced Alzheimer's disease is hospitalized for agitation. He responds to a calming robotic dog and agitation decreases

An 85-year-old recently widowed woman without family resources no longer complains of loneliness since having had a two-year-old rescue dog placed in her home by the local humane society

A 75-year-old man became suicidal after his wife of 45 years left him for another man. He states: 'What has prevented me from taking my life has been my devoted dog, Buddy. He gives me a reason to live.'

The above examples are just a few ways AAT has benefitted patients with a variety of presentations including depression, agitation, loneliness, and suicidality. These cases provide context and clarity for the research discussed in this chapter. The practice and utility of AAT in older adults demonstrates promising efficacy that merits future attention and further investigation.

References

Barker, S. B., Pandurangi, A. K., & Best, A. M. (2003). Effects of animal-assisted therapy on patients' anxiety, fear, and depression before ECT. *The Journal of ECT, 19*(1), 38–44.

Byers, A. L., & Yaffe, K. (2011). Depression and risk of developing dementia. *Nature Reviews Neurology, 7*(6), 323–331.

Chen, T.-T., Hsieh, T.-L., Chen, M.-L., Tseng, W.-T., Hung, C.-F., & Chen, C.-R. (2021). Animal-assisted therapy in middle-aged and older patients with schizophrenia: A randomized controlled trial. *Frontiers in Psychiatry, 12*, 713623. https://doi.org/10.3389/fpsyt.2021.713623

Cherniack, E. P., & Cherniack, A. R. (2014). The benefit of pets and animal-assisted therapy to the health of older individuals. *Current Gerontology and Geriatrics Research*, 623203. https://doi.org/10.1155/2014/623203, 2014.

Desai, A. K., Schwartz, L., & Grossberg, G. T. (2012). Behavioral disturbance in dementia. *Current Psychiatry Reports, 14*(4), 298–309. https://doi.org/10.1007/s11920-012-0288-5

Góngora Alonso, S., Hamrioui, S., de la Torre Díez, I., Motta Cruz, E., López-Coronado, M., & Franco, M. (2019). Social robots for people with aging and dementia: A systematic review of literature. *Telemedicine Journal and E-Health : The Official Journal of the American Telemedicine Association, 25*(7), 533–540. https://doi.org/10.1089/tmj.2018.0051

Jennings, A. A., Foley, T., Walsh, K. A., Coffey, A., Browne, J. P., & Bradley, C. P. (2018). General practitioners' knowledge, attitudes and experiences of managing behavioural and psychological symptoms of dementia: Protocol of a mixed methods systematic review and meta-ethnography. *Systematic Reviews, 7*(1), 62. https://doi.org/10.1186/s13643-018-0732-7

Majić, T., Gutzmann, H., Heinz, A., Lang, U. E., & Rapp, M. A. (2013). Animal-assisted therapy and agitation and depression in nursing home residents with dementia: A matched case-control trial. *The American Journal of Geriatric Psychiatry : Official Journal of the American Association for Geriatric Psychiatry, 21*(11), 1052–1059. https://doi.org/10.1016/j.jagp.2013.03.004

Marsilio, J., McKittrick, S., Garner, M., Umbell, L., Maiewski, S., & Wenos, J. (2018). Effects of a robotic cat on agitation and quality of life in individuals with dementia in a long- term care facility. *JMU Scholarly Commons*. https://commons.lib.jmu.edu/cgi/viewcontent.cgi?article=1045&context=pacapstones.

Moretti, F., De Ronchi, D., Bernabei, V., Marchetti, L., Ferrari, B., Forlani, C., Negretti, F., Sacchetti, C., & Atti, A. R. (2011). Pet therapy in elderly patients with mental illness. *Psychogeriatrics: The Official Journal of the Japanese Psychogeriatric Society, 11*(2), 125–129. https://doi.org/10.1111/j.1479-8301.2010.00329.x

Moyle, W., Cooke, M., Beattie, E., Jones, C., Klein, B., Cook, G., & Gray, C. (2013). Exploring the effect of companion robots on emotional expression in older adults with dementia: A pilot randomized controlled trial. *Journal of Gerontological Nursing, 39*(5), 46–53. https://doi.org/10.3928/00989134-20130313-03

Odendaal, J. S. J., & Meintjes, R. A. (2003). Neurophysiological correlates of affiliative behaviour between humans and dogs. *Veterinary Journal, 165*(3), 296–301.

Peluso, S., De Rosa, A., De Lucia, N., Antenora, A., Illario, M., Esposito, M., & De Michele, G. (2018). Animal-assisted therapy in elderly patients: Evidence and controversies in dementia and psychiatric disorders and future perspectives in other neurological diseases. *Journal of Geriatric Psychiatry and Neurology, 31*(3), 149–157. https://doi.org/10.1177/0891988718774634

Rodrigo-Claverol, M., Malla-Clua, B., Marquilles-Bonet, C., Sol, J., Jové-Naval, J., Sole-Pujol, M., & Ortega-Bravo, M. (2020). Animal-assisted therapy improves communication and mobility among institutionalized people with cognitive impairment. *International Journal of Environmental Research and Public Health, 17*(16). https://doi.org/10.3390/ijerph17165899

Schulman-Marcus, J., Mookherjee, S., Rice, L., & Lyubarova, R. (2019). New approaches for the treatment of delirium: A case for robotic pets. *The American Journal of Medicine, 132*(7), 781–782.

Smith, C., Bixler, D., George, A., Fusco, N., & DeLuca, A. (2020). A pilot study of animal assisted activity among hospitalized older adults. *Geriatric Nursing (New York, N.Y.), 41*(6), 905–908. https://doi.org/10.1016/j.gerinurse.2020.06.016

Waite, T. C., Hamilton, L., & O'Brien, W. (2018). A meta-analysis of Animal Assisted Interventions targeting pain, anxiety and distress in medical settings. *Complementary Therapies in Clinical Practice, 33*, 49–55. https://doi.org/10.1016/j.ctcp.2018.07.006

Wilson, R. S., Krueger, K. R., Arnold, S. E., Schneider, J. A., Kelly, J. F., Barnes, L. L., Tang, Y., & Bennett, D. A. (2007). Loneliness and risk of Alzheimer disease. *Archives of General Psychiatry, 64*(2), 234–240.

SECTION 3

Animal trainers' and patient perspectives

CHAPTER 12

Canine-assisted therapy for posttraumatic stress disorder in war veterans: perspective of an animal training organization

Jessica Nelson
Paws for Heroes, Houston, TX, United States

Mission, structure, and overview of Paws for Heroes

Paws for Heroes was founded in June 2013 in Houston, Texas. It was not originally the intention of its founders to create a new organization. Houston has the second largest veteran population in the United States (Mission United, 2022). There are numerous existing organizations in the Houston area providing a large variety of services to veterans—housing assistance, employment assistance, mental health counseling, and addiction recovery. However, the founders were particularly interested in veterans' services that involved the emotional support provided by dogs. The founders are dog lovers, who knew from personal experience the profound value of having a loving companion dog. The founders also had active-duty or veteran military members in their families, and they had seen the effects of posttraumatic stress disorder (PTSD) on those family members. Additionally, they were aware that Houston has a large animal overpopulation problem due to low rates of spay/neuter and the frequency of breeding during the warm weather for most of the year. Saving the lives of some of the dogs that ended up in shelters, subject to euthanasia, was an additional driver. Paws for Heroes decided to establish an organization that united veterans and companion dogs to give veterans a second chance at happiness and shelter dogs a second chance at life.

What Paws for Heroes discovered was several organizations that train service dogs for veterans. Service dogs do, of course, provide emotional

Animal Assisted Therapy Use Application by Condition
ISBN 978-0-323-98815-5
https://doi.org/10.1016/B978-0-323-98815-5.00014-8

support to their owners. However, a service dog, by definition, must also provide a service or "task" such as opening doors, detecting the onset of a seizure, or another physical service. More importantly, service dog training is a lengthy and very expensive proposition. The founders were surprised to find that there were no organizations in Houston providing dogs other than service dogs. "What about the veterans," they wondered, "who were physically whole, but could be helped by the emotional support of a dog?" Paws for Heroes was convinced that putting veterans together with emotional support companion dogs could make a difference in the lives of veterans with PTSD, feelings of isolation, depression, anxiety, or loneliness. Paws for Heroes believed they could provide excellent companion dogs to veterans without the lengthy wait for a service dog.

Paws for Heroes provides professionally trained emotional support companion dogs to military veterans. Paws for Heroes selects dogs from area shelters and rescue groups. Dogs are selected for their self-confidence, calmness, and owner orientation, among other positive traits. Both dog selection and training are conducted by experienced dog trainers who understand the mission of improving the lives of veterans suffering PTSD, military sexual trauma, and other military-related issues, while also saving the lives of homeless dogs. Paws for Heroes service is provided free of cost to each veteran. Paws for Heroes also stays in touch with clients for life. See Fig. 12.1.

Oscar and "Batman": This allows Paws for Heroes to ensure the veteran and their companion dog are doing well and to provide any assistance to the veteran if there are any issues with the dog after placed with the veteran. One Paws for Heroes client, Octavius, has attended several Paws for Heroes events and galas to speak to audiences about how "Petey" has helped improve Octavius' mental health and well-being and helped Octavius become less anxious. See Octavius and "Petey." Fig. 12.2.

Paws for Heroes veteran clients

Paws for Heroes' program is open to veterans of all eras and branches of service. Veterans served by Paws for Heroes program must reside in a house or apartment where a dog may live safely. While Paws for Heroes' services to find and train a dog are completely free, veterans must confirm that after

Figure 12.1 ***Oscar and "Batman."*** *(Photograph taken by Paws for Heroes volunteer. Written copyright permission on file.)*

the companion dog is placed with the veteran, the veteran has the financial ability to provide appropriate basic care (annual veterinary examinations, vaccinations, preventive medication, and healthful food) for their companion dogs in the future. If a veteran is not currently employed full time, Paws for Heroes requests veterans provide a financial summary documenting their monthly income and expenses. This information is helpful in determining whether the veteran will have sufficient discretionary income to care for a dog for the life of the animal as well as be able to afford medical emergencies that might occur.

Figure 12.2 ***Octavius and "Petey."*** *(Photograph taken by Paws for Heroes volunteer. Written copyright permission on file.)*

Application process

Paws for Heroes potential adopters are required to complete an adoption application. The application contains basic personal information (address, military service, household information, etc.) as well as information regarding any past ownership and care of dogs. The veteran is also required to submit a DD214 to demonstrate proof of military service.

Paws for Heroes conducts a multipart assessment of our applicant veterans. Our veterans must live in a pet-friendly home and must be able to physically take care of the dog. Paws for Heroes asks veterans to specify the size of dog they prefer. Paws for Heroes also inquires whether the veteran prefers a high, medium, or low-energy dog. However, Paws for Heroes also ensures that veteran's current lifestyle is compatible with the energy level that they have requested. If a veteran currently lives a sedentary lifestyle but indicates that they are interested in having a dog who would be a "running" buddy because the veteran intends to start a running program, Paws for Heroes would instead encourage the veteran to reconsider a dog that fits the veteran's current level of activity. Paws for Heroes also requires veterans read materials, which provide information on basic pet care, disease prevention, medication, nutrition, and substances that are toxic to dogs.

Ineligible veterans

Not all veterans who apply for a companion dog are eligible to receive a dog. Military members who are still active duty may not be able to take their companion dog with them to their new assignment. Therefore, Paws for Heroes does not place dogs with clients until the military member has separated from service. Veterans who are currently undergoing rehabilitation for mental health or substance abuse issues may not be eligible to receive a dog. The veteran must be in a stable environment physically, financially, and psychologically. If a veteran was recently homeless or displaced, Paws for Heroes would need to ensure that the veteran was in a safe and permanent living situation before placing a dog with the veteran. Paws for Heroes also does not place dogs in homes where children under the age of five reside (or visit the veteran regularly). Children under the age of five may not know how to behave appropriately with a dog. To ensure the safety of the children, Paws for Heroes only places dogs in homes with children who are older than five, so that the children can learn how to appropriately interact with a dog.

In-person interview meeting

Upon completion of the adoption application, Paws for Heroes meets with the veteran in person. This is an important part of the application and approval process. The in-person meeting is generally 45 minutes to an hour. This meeting allows the veteran to ask questions about the program.

The meeting also allows Paws for Heroes to get to know the veteran and their personality, energy level and history of PTSD, anxiety, and depression. Paws for Heroes believes it is important to establish trust and a rapport with a veteran before exploring personal questions about the veteran's mental health. Some veterans have a distrust of civilians or nonprofits that serve veterans.

Service animals versus companion animals

The initial meeting between the veteran and Paws for Heroes also allows Paws for Heroes to set expectations about the program. Paws for Heroes provides companion dogs or emotional support animals. Paws for Heroes emphasizes to the veteran that companion dogs are not legally classified as service dogs. Paws for Heroes explains the differences between companion dogs and service dogs and ensures that the veteran understands the differences. Paws for Heroes has observed that both in the veteran and civilian community, there is confusion and misinformation regarding service animals' vis-a-vis companion or emotional support animals. Paws for Heroes has found that it is helpful to reiterate this principle several times during the interview and approval process.

Paws for Heroes companion dogs are trained to Canine Good Citizen standard (Canine good citizen, 2022). Paws for Heroes dogs are trained for four to six weeks at a professional board and train facility that utilizes positive reinforcement training methods. Paws for Heroes dogs are obedience trained daily. Professional dog trainers "proof" the training by then taking the dogs to public spaces (such as Home Depot, PetSmart, etc.) where there are distractions and the dogs practice their commands (such as staying in a "down stay" while the trainer walks away and other people and dogs walk by). However, despite Paws for Heroes dogs being trained to ignore distractions and practicing commands in public, Paws for Heroes dogs do not have "public access" rights that are afforded to service animals. This means Paws for Heroes dogs cannot go into restaurants, stores, libraries, or other public places (ADA service animals revised 2010, 2010). Service dogs are trained for anywhere from 12 months to up to 24 months. Paws for Heroes mission is to place well-trained obedient dogs with veterans as soon as practicable after the dog has been trained.

The legal difference between a service animal and a companion animal

What Paws for Heroes explains to its potential clients is that a service dog is legally defined as a dog that has been individually trained to do work or perform tasks for an individual with a disability. The task(s) performed by the dog must be directly related to the person's disability. The dog must be trained to take a specific action when needed to assist the person with a disability. For example, a person with diabetes may have a dog that is trained to alert him when his blood sugar reaches high or low levels. Or a person who has epilepsy may have a dog that is trained to detect the onset of a seizure and then help the person remain safe during the seizure. Service animals are recognized under the Americans with Disabilities Act ("ADA") that protects individuals from being discriminated against on the basis of disability. The Department of Justice enforces the ADA. Public entities such as state and local government agencies, businesses, and nonprofit organizations (only covered entities) that provide goods or services to the public are required to make reasonable "accommodations" or "modifications" in their policies, practices, or procedures when necessary to accommodate people with disabilities (ADA, 2009).

The ADA makes a distinction between psychiatric service animals and emotional support animals. If the dog has been trained to sense that an anxiety attack is about to happen and take a specific action to help avoid the attack or lessen its impact, that would qualify as a service animal. However, if the dog's mere presence provides comfort, that would not be considered a service animal under the ADA. Consequently, service dogs are legally permitted to access to places that a companion dog is not. Under the ADA, a reasonable accommodation is allowing a service animal into an establishment that has a "NO PETS" policy. Therefore, for example, while a service dog would be allowed to accompany a veteran inside a restaurant, a Paws for Heroes companion dog would not be permitted (ADA service animals regs, 2010)[1]. More information about service animals and the ADA can be found here (Service animal resource hub, 2022).

[1] Service animals do not have to be trained by designated organizations. Individuals with disabilities have the right to train the dog themselves and are not required to use a professional service dog training program. Service animals are not required to wear a vest, ID tag, or specific harness under the ADA. Public entities are only allowed by law to ask the individual with a service animal two specific questions: (1) Is the dog a service animal required because of a disability? and (2) What work or task has the dog been trained to perform? Staff are not allowed to request any documentation for the dog, require that the dog demonstrate its task, or inquire about the nature of the person's disability. Public entities may not require "documentation" or "certification" or "papers," such as proof that the animal has been certified, trained, or licensed as a service animal, as a condition for entry of the service animal and their handler. There are individuals and organizations that sell service animal certification or registration documents online. These documents do not convey any rights under the ADA, and the Department of Justice does not recognize them as proof that the dog is a service animal. The ADA does not restrict the types of dog breeds that can be service animals. A service animal may not be excluded based on assumptions or stereotypes about the animal's breed or how the animal might behave. However, if a particular service animal behaves in a way that poses a direct threat to the health or safety of others, has a history of such behavior, or is not under the control of the handler, that animal may be excluded. If an animal is excluded for such reasons, staff must still offer their goods or services to the person without the animal present. If admitting service animals would fundamentally alter the nature of a service or program, service animals may be prohibited. In most settings, the presence of a service animal will not result in a fundamental alteration. However, there are some exceptions. For example, at a boarding school, service animals could be restricted from a specific area of a dormitory reserved specifically for students with allergies to dog dander. At a zoo, service animals can be restricted from areas where the animals on display are the natural prey or natural predators of dogs, where the presence of a dog would be disruptive, causing the displayed zoo animals to behave aggressively or become agitated. The ADA requires that service animals be under the control of the handler at all times. In most instances, the handler will be the individual with a disability or a third party who accompanies the individual with a disability. In the school (K-12) context and in similar settings, the school or similar entity may need to provide some assistance to enable a particular student to handle his or her service animal. The service animal must be harnessed, leashed, or tethered while in public places unless these devices interfere with the service animal's work or the person's disability prevents the use of these devices. In that case, the person must use voice, signal, or other effective means to maintain control of the animal. For example, a person who uses a wheelchair may use a long, retractable leash to allow her service animal to pick up or retrieve items. She may not allow the dog to wander away from her and must maintain control of the dog, even if it is retrieving an item at a distance from her. Or, a returning veteran who has PTSD and has great difficulty entering unfamiliar spaces may have a dog that is trained to enter a space, check to see that no threats are there, and come back and signal that it is safe to enter. The dog may be off-leash to do its job, but must be leashed at other times. Under control also means that a service animal should not be allowed to bark repeatedly in a lecture hall, theater, library, or other quiet place. However, if a dog barks just once, or barks because someone has provoked it, this will not mean that the dog is out of control. The ADA gives a person with a disability the right to be accompanied by his or her service animal to a food establishment, but these entities are not required to allow an animal to sit or be fed at the table. In addition, if a particular service animal is out of control and the handler does not take effective action to control it, or if it is not housebroken, that animal may be excluded.

In-home evaluation

After the initial interview with the veteran, Paws for Heroes will set up an in-home inspection. This in-home inspection is to ensure that the veteran's home is a safe and secure environment for a dog. If the veteran resides in a home with a fence, the fencing material and height should be adequate to keep a dog securely within the fence. Toxic plants (such as sago palms) should also be pointed to the veteran and removed.

There are occasions where Paws for Heroes visits a veteran's home and believes the home would not be conducive to having a dog reside there for safety or health concerns. In these instances, a letter is directed to the veteran after the home visit outlining that due to certain areas of concern in the veteran's home, the veteran has not been approved to adopt a dog from Paws for Heroes at this time. In the letter, Paws for Heroes outlines specific recommendations for the veteran to take to address the issues in the veteran's home and then the veteran can be reconsidered for approval.

Paws for Heroes' dog evaluation and placement process

Paws for Heroes adopts dogs from local shelters and rescue organizations. The shelters Paws for Heroes collaborates with do not euthanize for time or space. This means that if a dog is adopted out to the Paws for Heroes organization and the dog does not complete the training process, the dog can be returned to the shelter and will eventually be adopted out, not euthanized.

Paws for Heroes also works with local rescues. Paws for Heroes has created a summary outlining the qualifying and disqualifying criteria that Paws for Heroes uses with shelters and rescues to assist in identifying the type of dogs Paws for Heroes looks for in a candidate. Paws for Heroes does not search for dogs based upon the breed of the dog. Paws for Heroes instead focuses on the following: sociability (friendliness), temperament (calm), energy level (low), and confidence (curious and not easily startled).

Assessment process

Once a veteran has been approved to receive a dog, Paws for Heroes begins the search for a dog that is appropriate for that veteran. The search and selection process is the most critical—and time-consuming—element of the Paws for Heroes program. There are many good dogs, but only a small fraction of them have the right temperament and personality to be successful as emotional support animals for veterans suffering from PTSD, depression, and anxiety.

Paws for Heroes does not accept dogs into its program that are skittish or shy, have separation anxiety, thunderstorm anxiety or aggression, or other behavioral issues. While Paws for Heroes does not train dogs to become service animals, Paws for Heroes uses the same criteria to test dogs that service dog organizations utilize during evaluations. Paws for Heroes also follows a philosophy that the dog's current behavior is also likely its future behavior. If a dog is skittish or shy or is aggressive, Paws for Heroes does not accept the dog into the program with a plan to rehabilitate the dog to make it a more confident or a less aggressive dog. Volunteers who have experience working in shelters or with dog handling conduct the initial evaluation of a dog candidate. If the dog passes the initial evaluation, Paws for Heroes pays a professional dog trainer/behaviorist to return to the shelter and conduct a second evaluation on another occasion. This provides another opportunity to evaluate the dog's behavior as well as determine if the dog behaves and responds consistently with the first evaluation.

For the selection process, Paws for Heroes teams with a professional dog behaviorist with extensive experience. This professional dog behaviorist is a contractor who spends approximately 8 hours a week evaluating potential dogs. The behaviorist also meets with veterans both before and after matches are made, to provide guidance and additional training to the veteran. The behaviorist utilizes temperament and sociability testing to find a dog that is right for the veteran. Although Paws for Heroes dogs are companion animals, not service dogs, they must meet service dog criteria.

Paws for Heroes chooses each dog specifically for each veteran, but several characteristics are shared by all Paws for Heroes dogs. Any dog Paws for Heroes chooses must be between 2 and 4 years old—old enough to demonstrate its adult personality, but still young enough to have a long life with the veteran. It must be calm, confident, and friendly. A selected dog must not show any signs of aggression or anxiety, nor be easily startled or confused. Paws for Heroes looks for dogs who are affectionate and good with people and other animals, and who are owner-oriented. This means that the dog should not be aloof, or more interested in toys or other animals than its owner. Paws for Heroes may also seek a dog that can be desensitized to the veteran's individual situation, such as wheelchairs or scooters.

Paws for Heroes' testing/evaluation protocol

Paws for Heroes searches for potential candidate dogs at shelters and works with local rescue organizations to evaluate dogs. Paws for Heroes does not accept surrendered or donated dogs. This is because Paws for Heroes' mission is to adopt dogs that are either in shelters, rescues, or are homeless in its program.

When conducting searches at shelters, Paws for Heroes looks for shelters that have a sufficient number (at least 30 dogs) of adult dogs. This is because statistically, most dogs do not pass Paws for Heroes' evaluation process. Paws for Heroes tests for the following behaviors: (1) Dog's response to being approached in the kennel/dog run; (2) Dog's initial interaction with handler; (3) Dog's sociability/reaction to the handler after handler sits on a chair; (4) Dog's response/behavior to handler praising and petting the dog; (5) Dog's response to being stroked head to tail, having its paws handled/touched, having its teeth examined 3 times in a row, and being hugged; (6) Dog's response to playing tug of war with a squeaky toy/rope; (7) Guarding of dog food; (8) Guarding of a high value treat; and (9) Dog's startle and recovery to novel stimulus.

IMPORTANT: Only professional dog trainers and experienced volunteers should conduct shelter assessments of dogs. Paws for Heroes always conducts its shelter assessments and evaluations with two individuals: (1) a professional dog trainer and (2) experienced volunteers who have routinely handled dogs (someone who works at a dog shelter, a veterinary technician, etc.).

Dog Training Professional and Volunteer Roles: The volunteer should serve as the "handler" meaning this is the individual who holds the leashed dog and conducts the physical aspect of the assessment. The dog training professional should be in the room with the volunteer and dog candidate and be in close enough proximity to be able to evaluate the dog's response, reaction, and behaviors to the assessment, while still also remaining at a safe distance from the dog.

Dog's response to being approached in the kennel/dog run

This is an observation of the dog's body language and response to the handler's approach while the dog is the kennel or dog run. Paws for Heroes looks for dogs that are calm, appear friendly, make eye contact, and possibly wag their tail. If a dog is cowering in back of the run from fear, aggressively barking and lunging, snarling, or growling, staring at the handler, Paws for Heroes will not evaluate dogs exhibiting these types of behaviors. If the dog appears friendly, the handler will crouch down to the dog's level, turning sideways (so not directly facing the dog) to see how the dog reacts. Paws for Heroes is looking for the dog to continue to appear to be friendly and relaxed and potentially approach the handler (through a safe distance from the dog between the handler on one side and dog behind the door/gate).

Dog's initial interaction with handler

The purpose of this test is to determine the dog's initial level of sociability. When the dog is brought into the evaluation room, the dog should be leashed (4–6 foot leash) with a collar. The handler should take the leash, and without acknowledging or speaking to the dog, stand in the middle of the room for 1 minute while holding the dog's leash. The dog will likely sniff its environment and may approach the handler. Paws for Heroes is looking for a dog that stays near the handler, seeking to be petted and seeks further engagement. If the dog behaves in this manner, Paws for Heroes may rate the dog on its assessment form with a higher score. If the dog is lunging to try to get away from the handler or is out of control because it does not want to be leashed or controlled, Paws for Heroes may score the dog on the assessment form with a lower score.

If a dog is not able to be controlled or scores very low on this initial assessment, the dog should probably be returned to its kennel/dog run. Sociability is one of the most important criteria in the assessment and if a dog is not social with people, it doesn't matter how well the dog scores on all of the other assessments. Also, for safety reasons, if the dog is out of control or not social, the dog should not continue to be assessed since the remainder of the evaluation requires closer interactions between the dog and the handler.

Dog's sociability/reaction to the handler after handler sits on a chair

The purpose of this test is to further evaluate the dog's sociability while the handler is in seated position. While the dog is still leashed by the handler, the handler should sit down in a chair. The handler should not make eye contact or verbalize/speak to the dog or pet the dog. The handler should just sit down in the chair. Paws for Heroes is looking for a dog that seeks engagement with the handler (seeks affection, nudges the handler for attention, makes eye contact). Paws for Heroes may give higher scores to these types of behaviors. If a dog acts aloof and is uninterested in interacting with the handler, Paws for Heroes may assign a lower score to the dog.

Dog's response/behavior to handler praising and petting the dog

The purpose of this test is to evaluate the dog's reaction and behaviors once the handler begins to interact with the dog. While the handler is still sitting in the chair from the test in the prior assessment, the handler should begin gently petting the dog and praising the dog with an enthusiastic tone ("good girl/good boy!"). Paws for Heroes is looking for a dog that thoroughly enjoys this interaction and seeks more engagement with the handler, wagging its tail and become excited (but not too aroused or overexcited). Paws for Heroes may give higher scores to these types of behaviors. If the dog does not want to be petted or seeks to get away from the handler, Paws for Heroes may assign a lower score to the dog.

Stroking the dog head to tail

This is the first test that evaluates how the dog reacts to being handled. The handler should stand next to the dog leaning slightly over the dog's back, so that the handler is not near the dog's head/muzzle. Stroke the dog from the top of the head down to its tail and wait for seconds. Repeat stroking the dog two more times, waiting for seconds between each stroking. Paws for Heroes is looking for a dog that enjoys the interaction and petting and moves closer to the handler during each stroke. Paws for Heroes may give a higher score to these types of reactions. If the dog freezes or does not want to be touched or handled, Paws for Heroes may assign a lower score or terminate the examination for safety reasons if the dog appears to be uncomfortable being handled.

Handling the dog's paws

While still standing next to the dog, the handler will reach over the dog's back and lift up its rear paw. The handler should hold the paw and gently touch the dog's pads of the paw. Paws for Heroes is looking for a dog that will tolerate this type of body examination. Paws for Heroes may give a higher score to a dog that allows the examination and a lower score to a dog that does not tolerate the handling.

Examining the dog's teeth

While the leash is on the floor, and with one foot on the dog's leash (to control the dog and for safety), the handler should take one hand and hold the top of the dog's snout and with the other hand, lift the dog's lips to reveal its teeth. Repeat 3 times, waiting 5—10 seconds between each exam.

Do not proceed if the dog appears uncomfortable or reacts negatively after the first exam. Paws for Heroes is looking for a dog that will tolerate this exam. Paws for Heroes may assign a dog who tolerates the exam well a higher score; if the dog refuses to be examined or becomes too agitated during the exam, Paws for Heroes may assign a lower score.

Hugging the dog

If the handler is comfortable at this point in the evaluation process, proceed to have the handler hug the dog around its neck, with the handler's face pointing away from the dog's head. Paws for Heroes is looking for a dog that will tolerate this physical contact. If a dog tolerates this, Paws for Heroes may give the dog a higher score. If the dog's reaction is stiffness, or it appears uncomfortable, Paws for Heroes may assign a lower score.

Tug of war with squeaky toy/rope

The handler should take a squeaky toy or rope and engage the dog in a game of tug of war. The handler can pull on the toy/rope and encourage the dog to play. After the dog has become engaged for 15—20 seconds, the handler should cease tugging on toy but hold onto it and stand up as if to say, "the game is over." The dog should be assessed for its willingness to stop playing and looking toward the handler for what to do next. Paws for Heroes may give a dog that disengages quickly and calmly a higher score; a dog that cannot de-escalate after play or a dog who becomes aggressive in trying to grab the toy after the game is over may be assigned a lower score.

Food guarding and guarding high-value treats

This is the only part of the assessment that the professional dog trainer is physically involved in the exam. The professional dog trainer should be equipped with an "Assess a Hand" or similar tool. This will allow the trainer to stay within a safe distance from the dog while inserting what appears (to the dog) to be a human hand trying to take away the bowl of wet dog food. The handler should place a sufficient amount of wet dog food in a bowl, so that the dog can begin eating while the assessment is being conducted. The dog trainer should be holding the Assess a Hand and the handler should be standing directly opposite to the trainer, with the dog in between them. The handler should be holding the leash firmly with both hands and be in the ready position to pull the dog away from the dog trainer. Once the wet food is in the bowl, the bowl should be placed down in front of the dog by the trainer. Allow the dog to begin eating the food. After the dog takes several bites, but while there is still more than half of the dog food in the bowl, the dog trainer should begin gently stroking the dog with the Assess a Hand. The dog should be observed for freezing, growling, "punching" the food bowl, eating more quickly, etc. Next, the dog trainer should insert the Assess a Hand into the bowl while the dog is eating and begin to try push the dog's face out of the bowl. The dog should be observed for growling, eating more quickly, biting the hand, etc. Finally, the dog trainer should try to pull the bowl away from the dog and toward the trainer. It is important for the handler to have a very firm grip on the leash as the dog will be pulling strongly to continue eating the wet dog food. The dog should allow the bowl to be taken away without biting the hand, lunging, or growling. If the dog allows the bowl to be taken away, after a few seconds, the bowl can be returned to the dog and the dog can be allowed to finish eating the wet food. Paws for Heroes is looking for a dog that allows petting with the Assess a Hand and allows the Assess a Hand to be inserted into the bowl and take away the food without signs of biting, growling, etc.

Dog's startle and recovery to novel stimulus

The handler should have an umbrella that is the compact/spring opening type that can be fully opened by pushing one button. The handler should be approximately 4–6 feet away from the dog, and the dog should be facing toward the handler (and not distracted). The handler should push the button to open the umbrella in front of the dog and toss the umbrella near the dog, so that the umbrella lands near the dog. The dog should be observed for its reaction to the umbrella springing open and the dog's investigation of the umbrella after the umbrella is on the floor in front of it. Paws for Heroes may assign a higher score to a dog that recovers quickly from being startled and a dog that sniffs around or goes toward the umbrella out of curiosity. A dog that is unable to recover (runs and hides) or becomes very skittish around the open umbrella may be scored lower.

Veterinarian wellness exam

After Paws for Heroes accepts a dog into its program, the dog undergoes a veterinary wellness check. Dogs found to have irreversible health issues, such as hip dysplasia, would be separated from the Paws for Heroes program. Minor and curable health problems (including being heart-worm positive) are treated during foster homing and training. After the wellness check, the dog is microchipped, spayed or neutered (if not already), given up-to-date vaccinations, and supplied with preventive flea/tick and heartworm medications.

Foster process

Paws for Heroes' evaluation process at a shelter or rescue only provides part of the dog's personality. The most important part of the evaluation process is the foster period. If a dog passes both evaluations conducted by the volunteers and dog trainer, the dog is fostered for 4–6 weeks. The foster home is where the dog will learn house manners (no jumping on people, staying off of furniture, house training) and will provide more information about the dog's temperament, energy level, and personality. Behavioral issues will also sometimes emerge during the foster period. Anxiety issues such as thunderstorm anxiety or separation anxiety are incompatible with Paws for Heroes program.

Providing professional dog training

After foster care, the dog goes to a professional board-and-train facility, where it receives intensive professional training for 4–8 weeks. Paws for Heroes only engages professional dog trainers who use positive reinforcement. Paws for Heroes does not permit any type of negative reinforcement training methods. This means dog trainers cannot utilize prong collars, e-collars, shock collars, or any other type of methods.

Dogs are trained to obey basic commands—such as sitting, lying down, going to the crate, staying in place, "leave it," waiting for permission to exit an open door, and walking nicely by their owner on a loose leash.

Training is complete when the dog achieves the equivalent of the American Kennel Club's Good Canine Citizen standard. This standard evaluates dogs for positive behaviors such as accepting a friendly stranger, sitting politely for petting, and walking calmly through a crowd.

Presenting the dog to the veteran

Upon completion of the training, the dog is presented to the veteran. Paws for Heroes also provides the veteran a crate, crate liner, bed, leash, collar, grooming tools, toys, and bowls, as well as starter packages of preventive medications and nutritious food. Paws for Heroes is based in Houston, Texas. According to one 2019 study (Heartworm incidence survey, 2019), the southeast has the highest incidence of heartworms in the United States. Due to this fact, Paws for Heroes educates their veterans to understand what heartworm disease is, how it is contracted, and how it is prevented. As part of the adoption contract with Paws for Heroes, the veteran is required to provide authorization for Paws for Heroes to contact the veteran's veterinarian to obtain copies of heartworm medication purchased by the veteran to ensure the veteran is administering the heartworm medication monthly.

Providing additional training to the veteran and their emotional support dog

Paws for Heroes' commitment to the dog and the veteran is for life. Before the match is made, the veteran is required to visit the trainer's facility and learn how to execute the commands the dog has learned. This training is not conducted with the veteran's dog. This training is conducted with another unrelated dog that the trainer has been working with (or the trainer's own personal dog). During the month following the match, the veteran is required to meet with the trainer for weekly meetings, to confirm that the veteran is practicing the commands correctly, is able to interpret their dog's behavior, and to assure that the veteran and dog are settling in well together. Thereafter, Paws for Heroes remains in contact with the veteran by email, phone calls, and in-person home visits, to ensure that both the dog and the veteran are doing well.

"After-care" financial support

Paws for Heroes established an "after-care" program for their veterans. The reason for this program is that some of the dogs Paws for Heroes placed in subsequent years developed age-related health issues or diseases. Some of Paws for Heroes veterans are on fixed incomes. These veterans may not be in a financial position to afford expensive veterinary costs outside of annual visits and vaccinations. As a result, on a case-by-case basis, Paws for Heroes will provide veterans with financial assistance to fund the costs of veterinary care (e.g., veterinary specialists, prescription food for dogs that developed allergies, surgeries, etc.).

References

ADA. (2009). https://adata.org/learn-about-ada.
ADA service animals regs. (2010). www.ada.gov/regs2010/service_animal_qa.html.
ADA service animals revised 2010. (2010). www.ada.gov/service_animals_2010.htm.
Canine good citizen. (2022). www.akc.org/products-services/training-programs/canine-good-citizen/.
Heartworm incidence survey. (2019). www.avma.org/sites/default/files/2020-06/2020-08-01-story-06b-lg.jpg.
Mission United. (2022). www.unitedwayhouston.org/work/mission-united/#:~:text=The%20Greater%20Houston%20area%20is,population%20in%20the%20United%20States.
Service animal resource hub. (2022). https://adata.org/service-animal-resource-hub.

CHAPTER 13

Animal welfare

Chalotte Glintborg
Aalborg University, Department of Communication and Psychology, Aalborg, Denmark

Introduction

Animal-assisted interventions (AAIs) have seen a significant development in the last 50 years. Scientific research is beginning to provide initial evidence for the benefits of these interventions. However, ethical issues, particularly from the animals' point of view, are yet to be considered properly. This chapter addresses the ethical issues concerning the animals involved. As reported by Serpell et al. (2010), until recently, the general but unsubstantiated feeling across the field is that AAI represents "good" activities for animals to be engaged in and it is beneficial for humans. However, some concerns have been raised in an effort to observe AAI from the animal's viewpoint: are there any real benefits for the animal, and if so, what are they?

In terms of the animal's side, there is still "a void in the literature regarding the impact of these interventions on the therapy animals themselves" (Ng et al., 2015, pp. 357—376), even if over the years a few studies have begun to analyze some of the potential difficulties that can occur in the implementation of AAI. The major concerns relate to the animals' stress levels during these activities (that can lead to unwanted reactions and loss of motivation), the suitability of species and individual animals for involvement in AAI, their selection and training methods, their aging and retirement, the competence of the operators forming part of the multidisciplinary team, and the conditions of the AAI settings and associated risks (Bert et al., 2016; Evans & Gray, 2012; Fejsáskov et al., 2009; Iannuzzi & Rowan, 2015; Mongillo et al., 2015; Zamir, 2006).

Lack of animal awareness/ethics in AAT

The International Association of Human—Animal Interaction Organizations (IAHAIO White Paper. The IAHAIO definitions for Animal Assisted Intervention and guidelines for wellness of animals involved in AAI, 2018) is an umbrella group for about 100 organizations across the globe. Its

Animal Assisted Therapy Use Application by Condition
ISBN 978-0-323-98815-5
https://doi.org/10.1016/B978-0-323-98815-5.00013-6

member organizations have committed to a White Paper (IAHAIO White Paper. The IAHAIO definitions for Animal Assisted Intervention and guidelines for wellness of animals involved in AAI, 2018), which advocates the principles of One Health and One Welfare. This White Paper offers definitions and basic standards for human and animal welfare when providing AAIs. Under the superordinate category of AAIs, IAHAIO identifies four subordinate ones: animal-assisted therapy (AAT), animal-assisted education (AAE)/pedagogy, animal-assisted coaching, and animal-assisted activities. According to IAHAIO, great attention must be paid to the welfare of animals involved in AAI, given the fact that these interventions can be stressful for the animal (Evans & Gray, 2012; Fejsáskov et al., 2009; Iannuzzi & Rowan, 2015; Zamir, 2006; Ng et al., 2015, pp. 357−376). Companion animals involved in AAI are required to be present and actively participate in various situations and for multiple sessions in often short periods of time. To name just a couple of examples, in AAI settings, animals may have to deal with unfamiliar stimuli and relate to unknown patients/clients with a variety of disabilities or problems. This can be challenging and therefore stressful for the animal. An ethical reflection should accompany scientific research and the process of standardization of AAI to shape positive outcomes for both humans and animals in terms of welfare conditions.

Mental illness and animal-assisted interventions

A severe mental illness like posttraumatic stress disorder (PTSD) is known to have psychosocial consequences that can lead to a decreased quality of life. Research in AAT has revealed that the presence of a dog can have a positive effect on health, e.g., increase quality of life and lessen depression and anxiety. However, canine companionship is not a catch-all solution. Previous research has revealed methodological limitations that prohibit any clear conclusions, as well as a sparsity of critical reflection in anecdotal reports and case studies, which means that more research is needed to contextualize the findings.

Tedeschi et al. (2015, pp. 321−332) and Krause-Parello et al., (2016) have previously noted concerns that safety and animal welfare may not be adequately taken into consideration when the use of therapy and service animals expands into new fields, such as mental illnesses, including PTSD. There is also a lack of awareness of the need to integrate psychologically impactful interventions, including service dogs, with other interventions that are provided to support the same people by different means.

In mental health treatment, much effort in recent years has been put into developing an evidence base for which treatments work and which are questionable ("American Psychological Association Presidential Task Force on Evidence-Based Practice," 2006; Cusack et al., 2016) or even unintendedly harmful (Lilienfeld, 2007). There is a similar need to develop an evidence base in AAT (Fine et al., 2015, pp. 21–35; Kazdin, 2010) and in the use of animals to support therapy and rehabilitation. To achieve their potential benefit and avoid predictable pitfalls, the provision and use of animals for mentally ill clients must be informed by this evidence base. In the case of PTSD, programs have been developed with this need for integration in mind (Lefkowitz et al., 2005). The inclusion of expertise on animal welfare is also required, as is supervision of assistance dog providers, assistance dog recipients, and handlers in AAT.

Winkle et al. (2012) suggest that an assistance dog offers more than the tasks that it was trained to fulfill for the recipient. It also offers the opportunity of an emotional bond, and it may increase the recipient's autonomy, social inclusion, and participation. Kwong and Bartholomew (2011) found that a relationship often develops between an assistance dog and its owner that can be described as an attachment bond. In their interview study, most assistance dog recipients expressed strong grief after the death or retirement of their dog although they had received another. The issues of when an assistance dog should be retired and with whom it should spend its retirement years have been largely neglected in the literature (Ng & Fine, 2019).

Generally, the welfare of animals in AAI is a thorny issue. Legislation and requirements for certification and welfare of the animal vary across countries and states, and critiques have been raised of the validity of certification and of animal welfare issues in a growing AAI industry (Tedeschi et al., 2015, pp. 321–332). Some organizations advocate for standards, but the consensus around this and its impact may be less pervasive than is the case for interventions. Recommendations for intervention animals include ensuring that handlers have been educated in perceiving distress signals in the animals they work with, that the animal training is based on positive (reinforcement) techniques, and that each animal has limited working hours. In contrast, an assistance animal is "on call 24/7" (Tedeschi et al., 2015, pp. 321–332) for a recipient who may or may not have the resources or knowledge to attend to the animal's needs, notwithstanding their attachment to it. Assistance dog providers may traditionally have seen their task as being accomplished when the dog has been trained and handed over to the recipient, but interview studies suggest a need for assistance beyond that point (Gravrok et al., 2021; Ng & Fine, 2019).

Safety behavior

Safety-seeking behaviors are behaviors that are carried out (either overtly or covertly) in specific situations to prevent feared outcomes and is often seen in anxiety disorders like PTSD. Safety behaviors can include forms of avoidance, distraction, preparing, and checking. The feelings of safety induced by a dog can help adults with PTSD challenge their fears and can lead to a new evaluation of the world as relatively safe and of themselves as able and strong (Lefkowitz et al., 2005). However, an important caveat is the danger of the person misattributing causality to the dog, for instance, a veteran with PTSD using his assistance dog to sweep his apartment for bombs before entering it. Or as seen in a case study by Glintborg andand Hansen (Glintborg & Hansen, 2017) where a young woman, Helen, with PTSD, could go on walks and take baths when she was accompanied by her dog. In this way, she exposes herself to the fear of these things. However, she is only able to do this accompanied by the dog. Thus, without the dog, she is still not able to take a bath.

A review by Blakey and Abramowitz (2016, p. 13) stated that: "although safety behaviors are not unconditionally deleterious, they tend to interfere with exposure outcomes, possibly by promoting safety misattributions, preventing therapeutic information processing, or interfering with other mechanisms central to inhibitory learning theory" (Blakey & Abramowitz, 2016).

To avoid this misattribution, it is important that the use of animal assistance to deal with PTSD-related fear is done in close cooperation with a professional. Accordingly, the next step in therapy is, as happens with human supporters, that the animal assistance should be faded out progressively with each exposure session.

The treatment model suggested by Lefkowitz et al. (2005) called animal-assisted prolonged Exposure proposes a combination of AAT with exposure therapy. The rationale for the model was that clients would be more willing to put themselves in a feared situation if they were accompanied by a dog.

Coordinated AAT

Finally, another issue that has caused much concern in recent years is the fragmentation of services offered for mental health concerns in primary and volunteer-based sectors, the risk of these counteracting each other, and of important needs being overlooked in the false assumption that other entities dealt with them. The Convention on the Rights of Persons with Disabilities (Convention on the Rights of Persons with Disabilities, 2006) strongly advises that rehabilitation and treatment efforts are coordinated to ensure comprehensive rehabilitation based on multidisciplinary efforts in which professionals each use their area of expertise. In the area of AAT and the provision of AAT for the mentally ill, the relevant areas of expertise include professional knowledge of mental illness, client support and empowerment, animal training and welfare, human—animal psychology, and how to coordinate all these elements.

Thus, when working with AAI, all involved parties should be educated not only about the opportunities presented by animals but also on how to read the animal and secure its welfare. Furthermore, all providers involved need to cooperate closely, so that the animal becomes part of a professionally informed treatment plan rather than evoking parallel treatment processes that in the worst case may contradict each other. This leads us to a second concern: Is current knowledge and awareness sufficient?

Training methods and link between animal abuse and violence toward humans

A review by Ziv (2017) including 17 studies revealed that using aversive training methods (e.g., positive punishment and negative reinforcement) can endanger both the physical and mental health of dogs. Thus, those working with or handling dogs should rely on positive reinforcement methods and avoid using positive punishment and negative reinforcement as much as possible. This leads to some attention points in AAI in mental health.

For instance, anger and irritability are hyperarousal symptoms of PTSD. The Diagnostic and Statistical Manual of Mental Disorders (DSM-5) specifically lists anger as a common emotional reaction among people with PTSD.

In a similar vein, a substantial body of empirical evidence has increased the awareness of the cooccurrence of animal abuse and various forms of violence toward humans and other antisocial and criminal behaviors. Although the relationship is not a simple causal one, the correlation is reliable enough to suggest the importance of early intervention (Lunghofer & Shapiro, 2014). In response to these empirical findings, numerous policies have been implemented or are in development in the United States.

Canine body language

Not all canine body language is deliberate communication. Some language is reflexive in nature and reflects what is going on in the dog's autonomic nervous system. Stress signals can be deliberate communication, such as look away and lip lick, or they can be reflexive, such as a sudden hair loss and increased respiration.

Signals are a window into the dog's current emotional experience. Leaning these signals can help you answer various questions: Should I move closer to this dog? Should I or a client touch him/her? Is that dog feeling uncomfortable and therefore more likely to become reactive? Why is my dog unwilling to work in this setting? Knowing whether a dog is stressed, fearful, neutral, or friendly is of great importance when including animals in intervention.

Calming and negotiating signals

The term "calming signal" was first introduced by Turid Rugaas in her ground-braking book "*On Talking Terms With Dogs*" (Rugas, 2005). This book was the first book aimed at general readers that explained dog behaviors as deliberate communication. Turid Rugaas and Ståle Ødegård spent countless hours observing dogs' body language, taking note of body language signals used by dogs to communicate.

Body language had been studied in wolves and the term "cutoff signals" was used to describe body language to cut off aggression in other wolves (Aloff, 2018). However, Rugaas did not feel that the term "cutoff" was an appropriate term, as body language could be used as prevention—to avoid threats and to achieve a calming effect, negotiating rather than cutting off behavior.

Rugaas described how the body language signals could be used:

- as a means of the dogs calming themselves when feeling stressed or uncomfortable
- as an attempt to make others feel comfortable

- as an indication of good will and peaceful intentions within an interaction—a form of negotiation used in early stages of an interaction to avoid threats or enable dogs to make friends with others.

There are at least 30 signals or more, including subtle signals such as a head turn, shortening of the eye, turning the body away, lip lick, freezing, sitting, lying down, yawning, sniffing, and many more.

Communication and body language do not exist within a vacuum and should be viewed within context; it is important to observe the whole picture. The environment, circumstances, and parties involved give information as to what a dog may be communicating in that instance. Each dog is an individual and will have varying skills, experiences, and preferred ways of communicating. Some body language is deliberate, and some may be unconscious, just as you may unconsciously start to mess with your hair or scratch if you are feeling uncomfortable. Signals should not be looked at in isolation, but the whole body of the dog should be observed to ascertain what the dog might be communicating.

Dogs use deliberate signal in more than one way. Sometimes they use signals to create a calmer environment, but other times, they are trying to negotiate other things. In particular, controlling personal space is of vital importance to dogs. Dogs are constantly aware of and negotiating the use of their space.

For example, a dog enters a room filled with strangers. The dog repeatedly licks its lips, avoids eye contact, and sniffs obsessively. This dog is using signals to demonstrate that it is not threatening anybody, that it wishes it was not there and it is not comfortable. These are calming signals. This dog is trying to keep itself safe and wants the environment to calm down.

Obsessive sniffing is a vacuum activity (displacement behavior), a familiar behavior the dog can engage in to avoid focusing on what is causing it stress (that is, to calm itself down as well). Finally, the sniffing is also a sign of stress. Many of these behaviors are both deliberate communication and automatic reflection of the dog's internal state.

Dogs have a full emotional life, and status and possession are important to them. They can settle conflicts with tremendous force. At the same time, they are social animals, and this involves constant communication with others.

Dogs use this communication system toward us humans, simply because it is the language they know and think everyone understands. By failing to see your dog using calming signals on you, and perhaps even punish the dog for using them, you risk causing serious harm to your dog. Some may simply give up using the calming signals, including with other dogs. Others may get so desperate and frustrated that they get aggressive, agitated, or stressed out as a result.

Example

Henry calls for his dog Baloo. Henry has learned from friends that he needs to sound strict and dominant, so that Baloo will understand who is in charge. Baloo finds dad's voice to be aggressive and, being a dog, he instantly gives dad a calming signal in order to make him stop being aggressive. Baloo will perhaps lick his own nose, yawn, turn away—which will result in Henry becoming angry for real, because Henry perceives Baloo as being pig-headed, stubborn, and disobedient. Baloo is punished for using his calming signals to calm Henry.

This is a typical example of something that happens on an everyday basis with many dog owners. We need to learn to understand the language of dogs, so that we can understand what our dogs are telling us.

Examples of calming signals:

Yawning Fig. 13.1.

The dog may yawn when someone bends over him, when you sound angry, when there is yelling and quarreling in the family, when the dog is at the vet's office, when someone is walking directly at the dog, and when the dog is excited with happiness and anticipation—for instance, by the door

Figure 13.1 ***Yawning.*** In this picture, we see a dog yawning. The dog has just passed another dog.

when you are about to go for a walk, when you ask the dog to do something he does not feel like doing, when your training sessions are too long and the dog gets tired, when you have said no for doing something you disapprove of, and in many other situations.

Threatening signals (to walk straight at, reach for the dog, bending over the dog, staring into the dog's eyes, fast movements, and so on) will always cause the dog to use a calming signal.

All dogs know all the signals. When one dog yawns and turns his head to the side, the dog he is 'talking to" may lick his nose and turn his back—or do something completely different. The signals are international and universal. All dogs all over the world have the same language. A dog from Japan would be understood by a Norwegian elkhound. They will have no communication problems. However, humans are much worse than we claim to read and understand dogs—even our own pet dogs (Demirbas et al., 2016).

Licking Fig. 13.2.

Licking is another signal that is used often, especially for black dogs, dogs with a lot of hair around their faces, and others whose facial expressions for some reasons are more difficult to see than those of dogs with lighter colors, visible eyes, and long noses. But anyone can use licking, and all dogs understand it, no matter how quick it is. The quick little lick on the nose is easier to see if you watch the dog from in front. It is best seen if you can find somewhere you can sit in peace and quiet and observe. Once you have learned to see the lick, you will also be able to see it while walking the dog.

Figure 13.2 ***Licking.*** In this photo session, the dog is a bit uncomfortable. Thus, he is licking his lips.

Sometimes it is nothing more than a very quick lick, the tip of the tongue is barely visible outside the mouth, and only for a short second. But other dogs see it, understand it, and respond to it. Any signal is always returned with a signal.

Turning away/turning of the head Fig. 13.3.

The dog can turn its head slightly to one side, turn the head completely over to the side, or turn completely around, so that the back and tail are facing whoever the dog is calming. This is one of the signals you may see most of the time in dogs.

When someone is approaching your dog from its front, it will turn away in one of these ways. When you seem angry, aggressive, or threatening, you will also see one of these variations of the signal. When you bend over a dog to stroke it, it will turn its head away from you. When you make your training sessions too long or too difficult, he will turn his head away from you. When the dog is taken by surprise or takes someone by surprise, he will turn away quickly. The same happens when someone is staring or acting in a threatening way.

In most cases, this signal will make the other dog calm down, but what if the human does not see these signals and just continues? Often, we hear the saying "the dog just bit out of the blue." However, there are always a lot of small signals before the biting, but they have been overlooked. Dogs are experts at solving and avoiding conflicts with each other—they know how to deal with conflicts.

Play bow Fig. 13.4.

Going down with front legs in a bowing position can be an invitation to play if the dog is moving its legs from side to side in a playful manner. Just as

Figure 13.3 ***Turning away/turning head.*** Here we see three polar dogs. The dog in the middle is sending a calming signal to the others.

Figure 13.4 ***Play bow.*** The samoyed is inviting the malamute to play.

often, the dog is standing still while bowing and is using the signal to calm someone down. These signals often have double meanings and may be used in many ways—often the invitation to play is a calming signal by itself because the dog is making a potentially dangerous situation less tense and diverts with something safe.

When two dogs approach each other too abruptly, you will often see that they go into a play bow. This is one of the signals that is easy to see, especially because they remain standing in the bow position for a few seconds, so that you have plenty of time to observe it.

Sniffing the ground Fig. 13.5.

Sniffing the ground is a frequently used signal. In groups of puppies, you will see it a lot, and also when you and your dog are out walking and someone is coming toward you, in places where there is a lot going on, in noisy places, or when seeing objects that the dog is not sure of what is and finds intimidating.

Sniffing the ground may be anything from moving the nose swiftly down toward the ground and back up again—to sticking the nose to the ground and sniff persistently for several minutes.

Of course, dogs sniff a lot, also in order to 'read the paper" and enjoy themselves. Dogs are preprogrammed to use their noses, and it is their favorite activity. However, sometimes it is calming—it depends on the

Figure 13.5 ***Sniffing the ground.*** Two dogs and a cat are in a garden together. They use sniffing as a calming signal.

situation. So, pay attention to when and in which situations the sniffing occurs.

Walking slowly

High speed will be seen as threatening to many dogs, and they might want to go in to try and stop the one who is running. This is partly a hunting behavior and is triggered by the sight of a running human or dog. If the one running is coming straight at the dog, it involves a threat, and a defense mechanism sets in.

A dog who is insecure will move slowly. If you wish to make a dog feel safer, then you can move slower. When I see a dog react to me with a calming signal, I immediately respond by moving slower.

Is your dog coming very slowly when you call him? If so, check the tone of your voice—do you sound angry or strict? That may be enough for him/her to want to calm you down by walking slowly. Have you ever been angry with him/her when he comes to you? Then this may be why he does not trust you. Another reason to calm you may be if the dog is always put on a leash when called. Take a look at your dog the next time you call him. Does he give you any calming signals when coming? If he moves slowly, you may need to do something different in the way you act.

Freezing Fig. 13.6.

Figure 13.6 ***Freezing.*** The malamute has seen something in the field and is in a freeze position.

"Freezing"—is what we call it when the dog is stopping while standing completely still, sitting, or lying down and remaining in that position. This behavior is believed to have something to do with hunting behavior—when the prey is running, the dog attacks. Once the prey stops, the dog will stop too. We can often see this when dogs are chasing cats. This behavior, however, is used in several different situations. When you get angry and aggressive and appear threatening, the dog will often freeze and not move in order to make you be good again. Other times the dog may walk slowly, freeze, and then move slowly again. Many owners believe that they have very obedient dogs who are sitting, lying down, or standing completely still. Perhaps are they actually using calming signals? Very often, a dog will stop and remain calm when someone is approaching. If your dog wants to stop or move slowly in a situation like that, then let him. Also, if your dog is in a conflict situation with a human or dog, and is unable to escape, freezing may be one way to calm the other dog or person.

Sitting down/lifting one paw Fig. 13.7.

I have only rarely seen dogs lift their paw as a calming signal, but on a few occasions, it has clearly been used to calm another dog or human. To sit down, or an even stronger signal, to sit down with the back turned toward someone—for instance, the owner has a very calming effect. It is often seen when one dog wants to calm another dog who is approaching too quickly.

Figure 13.7 ***Lifting one paw.*** The samoyed has one paw lifted.

Dogs may sit down with their backs turned against the owner when he or she sounds too strict or angry.

Walking in curve Fig. 13.8.

This signal is frequently used as a calming signal, and it is the main reason why dogs may react so strongly toward meeting dogs when they are forced to walk straight at someone. Their instincts tell them that it is wrong to approach someone like that—the owner says differently. The dog gets

Figure 13.8 ***Walking in curve.*** The malamute is walking around the labrador in a curve to show kindness.

anxious and defensive. And we get a dog who is barking and lunging at other dogs, and eventually we have an aggressive dog.

Dogs, when given a chance, will walk in curves around each other. That is what they do when they meet off-leash and are free to do things their own way. Allow your dog to do the same when it is with you.

Some dogs need large curves, while others only need to walk slightly curved. Allow the dog to decide what feels right and safe for him/her. Let the dog walk in a curve when meeting an unfamiliar dog or person. Do not make it walk in a heel position while you are going straight forward—give him a chance to walk in a curve past the meeting dog. If you keep the leash loose and let the dog decide, you will often see that the dog chooses to walk away instead of getting hysterical.

For the same reason, do not walk directly toward a dog, but walk up to it in a curve. The more anxious or aggressive the dog is, the wider you make the curve.

By now you have learned about some of the more common calming signals. There are around 30 of them and many have yet to be described. I will mention a few more briefly, so that you can make further observations:

Smiling, either by pulling the corners of the mouth up and back, or by showing the teeth as in a grin.

Smacking the lips, wagging the tail—should a dog show signs of anxiety, calming, or anything that clearly has little to do with happiness, the wagging of the tail is not an expression of happiness, but rather that the dog wants to calm you.

Urinating on himself: A dog who is cowering and crawling toward his owner while wetting himself and waving his tail is showing three clear signs of calming and fear.

Wanting to get up into your face and *lick the corners of your mouth.*

Making the face round and smooth with ears close to the head in order to act like a puppy (No one will harm a puppy is what the dog believes).

Laying down with the belly against the ground: This has nothing to do with submission—submission is when the dog lays down with the belly up. Laying down with the belly toward the ground is a calming signal.

Some dogs act like puppies, jumping around and act silly, throwing sticks around, etc., if they discover a fearful situation. It is supposed to have, and does have, a calming effect.

Never force dogs into meeting others: Allow the dogs to use their language in meeting situations, so that they feel safe. Sometimes they will walk up to the other dog or human and just get along. Other times, they feel that it is safer to stay at a distance.

Suggested guidelines

An unequivocal set of ethical principles for AAI is still missing. In the search of these principles, especially regarding the animals involved, the Three Rs framework could be a source of valuable input for the regulation of the practice of AAI. The Three Rs principle proposed by Russell and Burch (1959) aimed to provide some reference points for ethical deliberation in animal research, with the goal to formulate a key strategy for the achievement of humane experimental techniques. It was aimed at encouraging ethical practices within the field of animal research. The three main guidelines of Russell and Burch's approach, that is, the Three Rs, are, briefly: to find a replacement for animals used in research, to reduce the number of animals used, and to refine experimental procedures and the animals' living conditions. That is, to reduce to an absolute minimum the amount of stress imposed on those animals that are still used.

Simonato et al. (2020) tried to adapt the Three Rs principle to the field of AAI for an ethical evaluation of the interventions, especially concerning the animal's side. The Three Rs tenet has a history of thorough study and discussion that has led to its current form, which allows for the initiation of a critical harm–benefit analysis in AAI. It allows us to consider, on the one hand, the benefits of the human–animal relationship and, on the other hand, the risks associated with the use we make of that relationship, which requires active (and potentially stressful) involvement of the animals.

Replacement

As already noted, Serpell et al. (2010), reported that the "general but unsubstantiated feeling across the industry was that AAI represents 'good' activities for animals to be engaged in. However, some concerns have been raised on the actual benefits. This reflection leads some authors to morally question the very involvement of animals in therapeutic and educational programs, which are beneficial to human health and well-being but may have negative effects on animal welfare (Marino, 2012). Whether the animal's involvement is necessary for a therapeutic effect in AAI has been recently addressed by Marino (2012). In a review of the recent literature, he concluded, consistently with other authors (Nimer & Lundahl, 2007; Souter & Miller, 2007), that the legitimate effects of AAT are likely to be moderate at best. Subsequently, for Marino, the question "How important is the animal in AAT?" remains unanswered.

Relative replacement has to do with the choice of the most appropriate species and the most appropriate individuals within the same species, based on their ethogram, their individual characteristics, their training, the aim of the project, the type of setting, and the category of patient/client involved. Concerning the choice of the most appropriate individual, we need to identify an animal who participates actively in the intervention, one not stressed by contact with humans and who possibly benefits from it. Therefore, we suggest—to paraphrase the original Russell and Burch statement—that relative replacement here requires us to look for animals with "higher socialization levels" instead of lower ones (Julius et al., 2014).

Reduction

When applied to animal research, reduction means lowering the number of animals used to the minimum necessary to obtain scientific results. In the case of AAI, the number of animals involved in the projects must first be in proper balance with the animals' workload. The decision to reduce the number of animals can be driven by difficulties in finding handlers and/or animals trained for a specific type of intervention, in combining their schedules, or by the need to lower costs. In any event, we suggest that four fundamental variables should be considered when considering whether the number of animals involved should be reduced: number of sessions, frequency of sessions, duration of sessions, and number of recipients. The right balance between these elements in relation to the project's objective protects the animal from being overburdened and consequent stress, allowing for the realization of a positive experience for both the human and nonhuman animal.

Refinement

In general terms, Simonato et al. (2020) identify three main spheres of action for refinement, which are intimately interrelated: environmental conditions, animal care, and people involved in the interventions. Regarding the environmental conditions, due attention should be paid to all the factors that can disrupt the intervention, such as sudden noises or annoying background noises, odors, and the presence of other animals (Ng et al., 2015, pp. 357–376). The characteristics of the facilities should also be taken into consideration: for example, facilities with drainable, soft, nonslip pavement that allow safe and protected activities, should be chosen as settings for AAI (Italian National Guidelines for Animal Assisted

Interventions, 2015). There should be escape routes for the animals, so that they can move away if they feel threatened or uncomfortable during the session (Fejsáskov et al., 2009), and generally animal-friendly logistics and equipment (e.g., harnesses) should be used. These arrangements should help prevent the risk of injury and discomfort for both animals and people involved in the interventions. Concerning the care of the animals, refinement applies to all spheres of life of those involved in AAI—before, during, and after each project and each session. The animals' health conditions should be constantly monitored through accurate clinical and behavioral assessments by the team's veterinarian and animal handler(s), using animal-friendly methods whenever possible (De Santis et al., 2017; Gehrke et al., 2011; Glenk, 2017; Glenk et al., 2014). A record sheet should be completed after every session to allow for the monitoring and assessment of the animal's condition, based on the assessment methods established by the team's veterinarian and shared with the animal handler. The record sheet should include data about the intervention (e.g., type of activities, features of the setting, number of patients/clients, handler's personal data, and length of sessions) and the animal's health and behavioral conditions, with regard to any physiological changes and signs of stress that may be observed during AAI sessions. Moreover, sessions should be stopped if animal health or welfare is perceived as compromised by the handler or the veterinarian.

We should also respect the animals' life stages and hormonal cycles. That is, puppies, females in estrous, animals in advanced stages of pregnancy, and lactating females should be excluded from AAI programs (Italian National Guidelines for AAI, 2015). Moreover, if an animal cannot be involved in AAI anymore, either due to its age or state of health, the animal must be guaranteed adequate living conditions and social interactions. Unfortunately, as reported in a few studies, many therapy animals are trapped in systems where they have little control over their social lives, and where they cannot avoid or escape unwanted social interactions. This can cause a decrease in their physical and mental well-being, which in turn has adverse effects on the success of the intervention (Fejsáskov et al., 2009; Serpell et al., 2010; Ng et al., 2015, pp. 357–376).

Particular attention should be paid to those animals who are held in the same facilities in which they "serve/work" and hence are defined as "residential animals." This condition, especially in facilities like nursing homes, long-term care facilities, psychiatric hospitals, and prisons, endangers the animal's well-being in terms of fatigue and possible burnout. Animals working in these conditions should be provided with appropriate time for rest in safe havens free of attention from the residents, as well as proper nutrition and exercise (Evans & Gray, 2012; Iannuzzi & Rowan, 2015; Serpell et al., 2010). AAI projects, but AAT and AAE, call for the involvement of a multidisciplinary team to effectively manage the intervention and make it beneficial to both parties. For example, the Italian Guidelines for AAI (Italian National Guidelines for Animal Assisted Interventions, 2015) provide that the team always includes at least a veterinarian, an animal handler, and another professional—a sanitary professional (AAT) or an educationalist (AAE), who is in charge of the client. This proposed structure is based on the so-called diamond model (Brooks, 2006), as opposed to the triangle model, in which the medical, psychological, or educational professional works without the assistance of an animal handler in the setting. Not all countries and AAI providers employ the diamond model; however, it is preferable to have that collaboration among the professionals on the two sides of the intervention, for welfare and safety purposes (Boggs et al., 2010; Schlote, 2009).

The fourth R: relationship

The relationship between "human animals" and "nonhuman animals" can entail sensitivity, harmony, happiness, and responsibility, or, on the contrary, manipulation, indifference, and cruelty (Nussbaum, 2007).

The feminist care tradition in animal ethics puts sympathy, empathy, and compassion at the core of relationships, calling for an effort by humans to listen to and communicate with nonhuman animals cognitively and emotionally (Donovan, 2006; Gruen, 2015).

It is equally important to stress the need for scientific knowledge in ethology and animal welfare, as they are essential to creating conditions for the animals to feel comfortable in the AAI setting. We need a good human–animal relation to achieve the therapeutic, educational, or socialization goals of AAI. So, we need to characterize the fourth R more clearly, starting with a classification of some types of human–animal relations as proposed by Zuolo (Zuolo, 2017). We can distinguish between exploitation, use, cooperation, and (individualized) relationships, which differ from each other based on the level of asymmetry between human and nonhuman animal interests (Zuolo, 2017, 2018). By "exploitation" is mean a purely instrumental relation, in which we use an animal to pursue a desired goal with total disregard of his/her welfare. "Use" is still an instrumental relation, but in this case, there is at least partial concern for the animal's welfare and natural features. "Individualized relationships are noninstrumental relations involving pet animals, in which species and individual needs are completely met. Finally, "cooperation," which is between "use" and "(individualized) relationships," is defined by Zuolo (Zuolo, 2017) as a "nonoccasional form of relation for the sake of producing an external good that does not harm the animal but rather benefits it" (p. 19). The notion of cooperation seems to apply rather well to the handler–animal relationship. If the animal and the handler must be "co-workers" (Evans & Gray, 2012), their bond needs to be well built.

Conclusions

In conclusion, in our relationship with animals in general and with AAI animals in particular, proper attention must be paid to each animal as an individual, with his/her own history, attitudes, and preferences. Doing so will allow us to deliver beneficial interventions for all the stakeholders and to develop animal ethics and welfare science, including the new field of AAI in the ethical analyses.

References

Aloff, B. (2018). *Canine body language. A photographical guide. Interpreting the native language of the domestic dog*. Dogwise Publishing.

American Psychological Association Presidential Task Force on Evidence-Based Practice. (2006). *American Psychologist, 61*, 271–285.

Bert, F., Gualano, M. R., Camussi, E., Pieve, G., Voglino, G., & Siliquini, R. (2016). Animal assisted intervention: A systematic review of benefits and risks. *European Journal of Integrative Medicine, 8*(5), 695–706. https://doi.org/10.1016/j.eujim.2016.05.005

Blakey, S. M., & Abramowitz, J. S. (2016). The effects of safety behaviors during exposure therapy for anxiety: Critical analysis from an inhibitory learning perspective. *Clinical Psychology Review, 49*, 1–15. https://doi.org/10.1016/j.cpr.2016.07.002

Boggs, J., Tedeschi, P., & Ascione, F. (2010). Animal-assisted approaches to child and adolescent mental health. In *Innovative interventions in child and adolescent mental health* (pp. 96–123). Routledge Taylor & Francis Group. https://doi.org/10.4324/9780203845004

Brooks, S. (2006). *Animal assisted psychotherapy and equine facilitated psychotherapy\nwith children who have trauma histories*. Guilford Press.

Convention on the Rights of Persons with Disabilities. (2006). http://www.refworld.org/docid/45f973632.html.

Cusack, K., Jonas, D. E., Forneris, C. A., Wines, C., Sonis, J., Middleton, J. C., Feltner, C., Brownley, K. A., Olmsted, K. R., Greenblatt, A., Weil, A., & Gaynes, B. N. (2016). Psychological treatments for adults with posttraumatic stress disorder: A systematic review and meta-analysis. *Clinical Psychology Review, 43*, 128–141. https://doi.org/10.1016/j.cpr.2015.10.003

De Santis, M., Contalbrigo, L., Borgi, M., Cirulli, F., Luzi, F., Redaelli, V., Stefani, A., Toson, M., Odore, R., Vercelli, C., Valle, E., & Farina, L. (2017). Equine assisted interventions (EAIs): Methodological considerations for stress assessment in horses. *Veterinary Sciences, 4*(3). https://doi.org/10.3390/vetsci4030044

Demirbas, Y. S., Ozturk, H., Emre, B., Kockaya, M., Ozvardar, T., & Scott, A. (2016). Adults' ability to interpret canine body language during a dog—child interaction. *Anthrozoös, 29*(4), 581—596. https://doi.org/10.1080/08927936.2016.1228750

Donovan, J. (2006). Feminism and the treatment of animals: From care to dialogue. *Signs: Journal of Women in Culture and Society, 31*(2), 305—329. https://doi.org/10.1086/491750

Evans, N., & Gray, C. (2012). The practice and ethics of animal-assisted therapy with children and young people: Is it enough that we dont eat our co-workers? *British Journal of Social Work, 42*(4), 600—617. https://doi.org/10.1093/bjsw/bcr091

Fejsáskov, Kottferov, Marekov, J., Jakuba, T., Ondrašovičov, O., & Ondrašovič, M. (2009). Ethical aspects related to involvement of animals in animal assisted therapy. *Folia Veterinaria, 53*(1), 62—64.

Fine, A. H., Tedeschi, P., & Elvove, E. (2015). *Forward thinking*. Elsevier BV. https://doi.org/10.1016/b978-0-12-801292-5.00003-1

Gehrke, E. K., Baldwin, A., & Schiltz, P. M. (2011). Heart rate variability in horses engaged in equine-assisted activities. *Journal of Equine Veterinary Science, 31*(2), 78—84. https://doi.org/10.1016/j.jevs.2010.12.007

Glenk, L. M. (2017). Current perspectives on therapy dog welfare in animal-assisted interventions. *Animals, 7*(2). https://doi.org/10.3390/ani7020007

Glenk, L. M., Kothgassner, O. D., Stetina, B. U., Palme, R., Kepplinger, B., & Baran, H. (2014). Salivary cortisol and behavior in therapy dogs during animal-assisted interventions: A pilot study. *Journal of Veterinary Behavior: Clinical Applications and Research, 9*(3), 98—106. https://doi.org/10.1016/j.jveb.2014.02.005

Glintborg, C., & Hansen, T. G. B. (2017). How are service dogs for adults with post traumatic stress disorder integrated with rehabilitation in Denmark? A case study. *Animals, 7*(5). https://doi.org/10.3390/ani7050033

Gravrok, J., Bendrups, D., Howell, T., & Bennett, P. (2021). The experience of acquiring an assistance dog: Examination of the transition process for first-time handlers. *Disability and Rehabilitation, 43*(1), 49—59. https://doi.org/10.1080/09638288.2019.1615561

Gruen, L. (2015). *Entangled empathy: An alternative ethic for our relationship with\nAnimals*. Lantern Books.

Iannuzzi, D., & Rowan, A. N. (2015). Ethical issues in animal-assisted therapy programs. *Anthrozoös, 4*(3), 154—163. https://doi.org/10.2752/089279391787057116

IAHAIO White Paper. The IAHAIO definitions for Animal Assisted Intervention and guidelines for wellness of animals involved in AAI. (2018). http://iahaio.org/best-practice/white -paper-on-animal-assisted-interventions/.

Italian National Guidelines for Animal Assisted Interventions. (2015). http://www.salute.gov.it/imgs/C_17_opuscoliPoster_276_allegato.pdf.

Julius, H., Beetz, A., Kotrschal, K., Turner, D., & s-Moberg, Uvn (2014). *L'attaccamento agli animali. Una visione integrata della relazione uomo-animale nella pet therapy [Attachment to pets: An integrative view of human-animal relationship with implications for therapeutic practice]*. Hogrefe Editore.

Kazdin, A. E. (2010). Methodological standards and strategies for establishing the evidence base of animal-assisted therapies. In *Handbook on animal-assisted therapy* (pp. 519—546). Elsevier Inc. https://doi.org/10.1016/B978-0-12-381453-1.10025-X

Kwong, M., & Bartholomew, K. (2011). "Not just a dog": an attachment perspective on relationships with assistance dogs. *Attachment & Human Development, 13*(5), 421—436. https://doi.org/10.1080/14616734.2011.584410

Krause-Parello, C. A., Sarni, S., & Padden, E. (2016). Military veterans and canine assistance for post-traumatic stress disorder: A narrative review of the literature. *Nurse Education Today, 47*, 43–50. https://doi.org/10.1016/j.nedt.2016.04.020

Lefkowitz, C., Paharia, I., Prout, M., Debiak, D., & Bleiberg, J. (2005). Animal-assisted prolonged exposure: A treatment for survivors of sexual assault suffering post-traumatic stress disorder. *Society and Animals, 13*(4), 275–295. https://doi.org/10.1163/1568530057746553654

Lilienfeld, S. O. (2007). Psychological treatments that cause harm. *Perspectives on Psychological Science, 2*(1), 53–70. https://doi.org/10.1111/j.1745-6916.2007.00029.x

Lunghofer, L., & Shapiro, K. (2014). The co-occurrence of human violence and animal abuse: Policy implications and interventions. *Psyke & Logos, 35*, 130–135.

Marino. (2012). Construct validity of animal-assisted therapy and activities: How important is the animal in AAT? *Anthrozoös: A Multidisciplinary Journal of the Interactions of. People & Animals, 25*, 139–151. https://doi.org/10.2752/175303712X13353430377219

Mongillo, P., Pitteri, E., Adamelli, S., Bonichini, S., Farina, L., & Marinelli, L. (2015). Validation of a selection protocol of dogs involved in animal-assisted intervention. *Journal of Veterinary Behavior, 10*(2), 103–110. https://doi.org/10.1016/j.jveb.2014.11.005

Ng, Zenithson, Albright, J., Fine, A. H., & Peralta, J. (2015). *Our ethical and moral responsibility*. Elsevier BV. https://doi.org/10.1016/b978-0-12-801292-5.00026-2

Ng, Z., & Fine, A. (2019). Paving the path toward retirement for assistance animals: Transitioning lives. *Frontiers in Veterinary Science, 6*. https://doi.org/10.3389/fvets.2019.00039

Nimer, J., & Lundahl, B. (2007). Animal-assisted therapy: A meta-analysis. *Anthrozoos, 20*(3), 225–238. https://doi.org/10.2752/089279307X224773

Nussbaum, M. C. (2007). *Frontiers of justice: Disability, nationality, species membership*. Belknap Press.

Rugas, T. (2005). *On talking terms with your dog: Calming signals*. Canada.

Russell, W. M., & Burch, R. L. (1959). *The principles of humane experimental technique*.

Schlote, S. M. (2009). *Animal-assisted therapy and equine-assisted therapy/learning in Canada: Surveying the current state of the field, its practitioners, and its practices (Doctoral dissertation)*.

Serpell, J. A., Coppinger, R., Fine, A. H., & Peralta, J. M. (2010). Welfare considerations in therapy and assistance animals. In *Handbook on animal-assisted therapy* (pp. 481–503). Elsevier Inc. https://doi.org/10.1016/B978-0-12-381453-1.10023-6

Simonato, M., de Santis, M., Contalbrigo, L., de Mori, B., Ravarotto, L., & Farina, L. (2020). The three R's as a framework for considering the ethics of animal assisted interventions. *Society and Animals, 28*(4), 395–419. https://doi.org/10.1163/15685306-00001767

Souter, M. A., & Miller, M. D. (2007). Do animal-assisted activities effectively treat depression? A meta-analysis. *Anthrozoos, 20*(2), 167–180. https://doi.org/10.2752/175303707X207954

Tedeschi, P., Pearson, J. A., Bayly, D., & Fine, A. H. (2015). *On call 24/7—The emerging roles of service and support animals*. Elsevier BV. https://doi.org/10.1016/b978-0-12-801292-5.00023-7

Winkle, M., Crowe, T. K., & Hendrix, I. (2012). Service dogs and people with physical disabilities partnerships: A systematic review. *Occupational Therapy International, 19*(1), 54–66. https://doi.org/10.1002/oti.323

Zamir, T. (2006). The effects of using aversive training methods in dogs - a review. *Journal of Veterinary Behavior, 14*(2), 50–60.

Ziv, G. (2017). The effects of using aversive training methods in dogs—a review. *Journal of Veterinary Behavior, 19*, 50–60. https://doi.org/10.1016/j.jveb.2017.02.004

Zuolo. (2017). *Human-animal relations and the idea of cooperation (working paper- LPF n. 1)*. Torino, Italia: Centro Einaudi—Laboratorio di Politica Comparata Filosofia Pubblica.
Zuolo, F. (2018). *Etica e animali. Come è giusto trattarli e perché.*

CHAPTER 14

Canine-assisted therapy for posttraumatic stress disorder: a war veteran's perspective

Paws for Heroes Client - Army Veteran
Houston, TX, United States

I was born and raised in Texas. I married my high school sweetheart, and we started our own life. We wanted to see the world and wanted to travel. But after 9/11, I wanted to do something to protect the country. I also wanted the financially stability the military offered. I joined the Army in 2006. I spent the next 6 years serving. I spent time in Alaska, Virginia, and South Carolina. I was deployed to Iraq, my tour split by Operation Iraqi Freedom and Operation Enduring Freedom from 2008 to 2009.

I had the opportunity to support Special Forces and absolutely loved it during my time in Iraq. Special Forces treat you with every bit of respect—it didn't matter what rank you were. After returning home and eventually going through the med board process, I was honorably discharged and was medically retired with 100% disability rating as a result of having surgery on my hand and on my elbow (due to an injury in Iraq), and posttraumatic stress disorder (PTSD).

After I separated from the Army, things between me and my wife were a little bit difficult for a while. My mindset for the first two months was to do absolutely nothing. And that's what I did for a while. I didn't even sleep in our bed most nights. I would go for walks at night and fall asleep at the pool in the apartment complex. My wife would get pissed off because I would expect her to come out and find me (which was not fair to her).

I got a job at an aerospace engineering company, and I was working a lot. Some nights I didn't get home until after 11 p.m. in the evening. My wife was worried about me while also trying to take care of our two sons who were 10 and 12 years old at the time. They, of course, didn't understand anything about my PTSD. There are also several months where I don't remember seeing my kids, I don't remember talking to them. I don't remember anything. There were times I didn't leave my bed for what felt like years....

Animal Assisted Therapy Use Application by Condition
ISBN 978-0-323-98815-5
https://doi.org/10.1016/B978-0-323-98815-5.00012-4

My wife was the one who looked up Paws for Heroes about a companion dog. She was looking for something to help me. I wasn't interested in going to a therapist, and I was in a very dark hole. My kids weren't a priority, my wife wasn't a priority. I thought a dog who was trained might be able to help. Paws for Heroes found Bear, a stray dog who was picked up by animal control and brought to a high kill shelter. Bear's foster, a volunteer at the shelter, reached out to Paws for Heroes to tell them about this calm, sweet and very sociable dog. Paws for Heroes evaluated Bear and accepted him into their program.

The first day I received Bear, we watched the Super Bowl together. That night, he slept between my legs with his head on my thigh. It was comfortable and it was calming. After Bear was placed with me, I've actually been able to go to college, I can do really anything that I want to. He loves people. Sometimes he can sense when I am getting anxious, and he will lean his body up against me. It's something that brings me to the present and keeps me out of "my head."

I've had Bear for 6 years. A lot of people wonder what do companion dogs do for veterans like me. To be honest, the main thing with Bear is that he is really not asking anything from you, so it's not an additional load on your plate. He literally just wants to be with you. When your family members might be angry at you because you won't do something or you are upset with them (especially with teenagers!) you have this happy dog that just wants to be petted and shows you pure happiness.

After I got Bear, I was slowly able to start getting out of the house with my family. We went to a basketball game. That would have never happened without Bear. Another thing Bear does is help distract me when I am in public. Normally, I would be focusing on all of the entries and exits indoors. However, when Bear is with me, while I might still be looking for exits/entries, but I am not obsessed with it. Bear has a calming presence for me. Has Bear eliminated my PTSD? No, and Paws for Heroes didn't represent that he would. Bear helps me. Bear also helps my family. He loves my wife and my kids, and they love him. Bear has brought all of us closer together too.

One of the biggest things Bear has done for me is orient me when I have night terrors. It happens most frequently after I have watched military war movies. I would wake up in the middle of the night, and I don't recognize my wife and I am dreaming I am back in Iraq. Bear would jump on the bed and stand between me and my wife. Even though my wife was talking to me, I wouldn't be able to recognize her. It takes a few seconds, but eventually after seeing Bear, I would realize I was home.

I wouldn't change my decision to serve in the Army. I was proud to serve our country and am grateful that all of my children have the ability to go to college and they don't have the burden of paying for it. I am especially appreciative that my wife reached out to Paws for Heroes and that we have these types of organizations that can offer this type of assistance to veterans. Since Bear also came from a high kill shelter, I feel like we have an unspoken understanding that not only did he save me but also we were able to save him. Bear has become a member of the family, and we have become members of Paws for Heroes' family.

SECTION 4

Conclusion

CHAPTER 15

Conclusion and future studies and applications of animal-assisted therapy

Eric Altschuler[1,2]
[1]Department of Physical Medicine and Rehabilitation, Metropolitan Hospital, New York, NY, United States; [2]Department of Rehabilitation Medicine, New York Medical College, Valhalla, NY, United States

To conclude, we return to Kate's soliloquy about the malady afflicting her husband, war veteran Hotspur, from Shakespeare's Henry IV, Part 1 (Act 2, Scene 3). Using our modern-day war veteran's account (from the previous chapter), we see how the veteran's emotional support dog "Bear" helps to alleviate the signs and symptoms of posttraumatic stress disorder (PTSD) ((Shay, 1994), noted in square brackets) so presciently described by Shakespeare:

O, my good lord, why are you thus alone? [social withdrawal] A
For what offence have I this fortnight been [random rage at family] B
A banish'd woman from my Harry's bed? [sexual disinterest/disfunction] C
Tell me, sweet lord, what is't that takes from thee
Thy stomach, pleasure and thy golden sleep? [loss of interest, insomnia] D
Why dost thou bend thine eyes upon the earth, [depression] E
And start so often when thou sit'st alone? [hyperactive/startle reaction] F
Why hast thou lost the fresh blood in thy cheeks;
And given my treasures and my rights of thee
To thick-eyed musing and curst melancholy?
In thy faint slumbers I by thee have watch'd, [fragmented, vigilant sleep] G
And heard thee murmur tales of iron wars; [traumatic dreams, reliving episodes of combat] H
Speak terms of manage [horsemanship] to thy bounding steed;
Cry 'Courage! to the field!' And thou hast talk'd
Of sallies and retires, of trenches, tents,
Of palisadoes, frontiers, parapets,
Of basilisks, of cannon, culverin,
Of prisoners' ransom and of soldiers slain,

Animal Assisted Therapy Use Application by Condition
ISBN 978-0-323-98815-5
https://doi.org/10.1016/B978-0-323-98815-5.00011-2

And all the currents of a heady fight.
Thy spirit within thee hath been so at war,
And thus hath so bestirr'd thee in thy sleep,
That beads of sweat have stood upon thy brow [night sweats] I
Like bubbles in a late-disturbed stream;
And in thy face strange motions have appear'd,
Such as we see when men restrain their breath
On some great sudden hest. O, what portents are these?
Some heavy business hath my lord in hand,
And I must know it, else he loves me not.

A—After Bear was placed with me, I've actually been able to go to college, and I can do really anything that I want to.

After I got Bear, I was slowly able to start getting out of the house with my family. We went to a basketball game. That would have never happened without Bear.

B—When your family members might be angry at you because you won't do something or you are upset with them (especially with teenagers!), you have this happy dog that just wants to be petted and shows you pure happiness.

C—I would wake up in the middle of the night, and I don't recognize my wife and I am dreaming I am back in Iraq. Bear would jump on the bed and stand between me and my wife. Even though my wife was talking to me, I wouldn't be able to recognize her. It takes a few seconds, but eventually after seeing Bear, I would realize I was home.

D—The first day I received Bear, we watched the Super Bowl together. That night, he slept between my legs with his head on my thigh. It was comfortable and it was calming.

E—Sometimes, he can sense when I am getting anxious and he will lean his body up against me. It's something that brings me to the present and keeps me out of "my head."

F—Another thing Bear does is help distract me when I am in public. Normally, I would be focusing on all of the entries and exits indoors. However, when Bear is with me, while I might still be looking for exits/entries, but I am not obsessed with it. Bear has a calming presence for me.

G–I—One of the biggest things Bear has done for me is orient me when I have night terrors. It happens most frequently after I have watched military war movies. I would wake up in the middle of the night, and I don't recognize my wife and I am dreaming I am back in Iraq. Bear would jump on the bed and stand between me and my wife.

Our war veteran concludes: "Has Bear eliminated my PTSD? No, and Paws for Heroes didn't represent that he would. Bear helps me. Bear also helps my family. He loves my wife and my kids, and they love him. Bear has brought all of us closer together too."

Animal-assisted therapy (AAT) has been very helpful for our war veteran with PTSD. Now further research of AAT is warranted.

More research on the physiologic effects of AAT and best methods and approaches of incorporating AAT into psychotherapy is needed. Future investigations should seek to identify the best approaches and further study the efficacy of using canine, equine, and avian-assisted therapy for PTSD. There needs to be more research on AAT for sexual trauma survivors and individuals with PTSD secondary to other causes of trauma. The incidence and prevalence of PTSD in sexual trauma survivors need to be studied. The mechanism of action of AAT for PTSD should be considered for study.

More research on the uses of AAT for pediatric and geriatric patients, and AAT for heart failure and other patients with cardiac and general medical conditions is important.

The ethical practice of AAT is essential, and further research on animal welfare is crucial as is the development of clinical guidelines and standards for AAT.

Finally, a search for examples of PTSD induced by trauma of subordinates (Altschuler, 2016) in history, literature, and art is warranted as well as consideration of societal implications of such PTSD as an occupational hazard in current and future military commanders and civilian leaders.

References

Altschuler, E. (2016). PTSD induced by the trauma of subordinates: The Robert Gates syndrome. *Occupational Medicine, 66*(3), 182. https://doi.org/10.1093/occmed/kqv201

Shay, J. (1994). *Achilles in vietnam.* Scribner.

Index

Note: 'Page number followed by "f" indicate figure and "t" indicate table.'

A

"After-care" financial support, 189
Alcohol use disorder (AUD), 76
Alzheimer's disease, 91
Americans with Disabilities Act (ADA), 49
Animal-assisted interventions (AAIs), heart failure. *See also* Animal welfare
  cardiac outcomes, 148–149, 158–159
  exemplar, 155–158
  heart rate modulation, 149–151
  physical activity, 153–154
  stress mitigation, 151–153
Animal welfare
  animal abuse and violence, 195–196
  animal awareness/ethics, 191–192
  canine body language
    calming and negotiating signals, 196–198, 198f–202f
    walking slowly, 202–205, 203f–204f
  guidelines, 206–210
  mental illness, 192–195
    coordinated AAT, 195
    safety behavior, 194
Attachment theory, 58–59
Autism spectrum disorders (ASDs)
  dogs, 124–125
  horses, 134, 135f
Avian-assisted therapy
  Alzheimer's disease, 91
  environment, 93–95
  establishment, 92–93, 94t
  parrot, 85–88, 86f
  patient history, 88–91, 90f

C

Calming signals, 196–198, 198f–202f
Canine-assisted ambulation (CAA) program, 156–157
Canine-assisted therapy, 10, 12–13
  Paws for Heroes
    "after-care" financial support, 189
    application process, 176–177
    assessment process, 181–187
    emotional support dog, 189
    evaluation and placement process, 181
    foster process, 188
    ineligible veterans, 177
    in-home evaluation, 181
    in-person interview meeting, 177–178
    mission, structure, and overview, 173–174, 175f
    presentation, 188–189
    professional dog training, 188
    service animals *vs.* companion animals, 178–180
    veteran clients, 174–175
    veterinarian wellness exam, 187–188
  veteran's perspective, 215–217
Canine body language
  calming and negotiating signals, 196–198, 198f–202f
  walking slowly, 202–205, 203f–204f
Cerebral palsy (CP), 131–134
Cognitive processing therapy (CPT), 22
COVID-19, 92

D

Dogs
  autism spectrum disorders (ASDs), 124–125
  hospitalized children, 120–121
  mental health disorders, 121–124

Dogs (*Continued*)
reading disorders, 128–129
severe disabilities, 125–128, 128f

E

Elephants, 136
Equine-assisted therapy (EAT)
anxiety, depression, and anger, 78
diagnostic criteria, PTSD, 66–67
ethical practices, 73
experiential learning, 72–73
family considerations, 74–76
focus, concentration, and memory, 77
independence and self-mastery, 77
military culture and treatment, 66
prevalence, 65–66
relationships, 78
substance abuse disorder, 76
symptoms of PTSD, 67–69
treatment, 69–72
with veterans, 73–74
Equine-assisted therapy for PTSD (EAT-PTSD), 20–21
Equine-facilitated cognitive processing therapy (EF-CPT), 24
Eye movement desensitization and reprocessing (EMDR), 22

F

Fetal alcohol syndrome (FAS), 126–127
Freezing, 202–203

G

Geriatric patients
complications, 167
depression and anxiety, 166
indications, 165
patient history, 167–168
psychiatric illnesses in older adults, 166
risk factors, dementia, 165
Guinea pigs, 135–136

H

Heart disease, 147
Heart failure
animal-assisted interventions
cardiac outcomes, 148–149, 158–159
exemplar, 155–158
heart rate modulation, 149–151
physical activity, 153–154
stress mitigation, 151–153
pathophysiology and management, 147–148
prevalence, 147
Hippotherapy (HPOT), 70
Horses
autism spectrum disorders (ASDs), 134, 135f
cerebral palsy, 131–134
mental health, 130–131

M

Mental health disorders
dogs, 121–124
horses, 130–131
Mental illness, 192–195
coordinated AAT, 195
safety behavior, 194
Military sexual trauma (MST), 101

N

Negotiating signals, 196–198, 198f–202f

P

Parrot, 85–88, 86f
Paws for Heroes' program
"after-care" financial support, 189
application process, 176–177
assessment process, 181–187
emotional support dog, 189
evaluation and placement process, 181
foster process, 188
ineligible veterans, 177
in-home evaluation, 181
in-person interview meeting, 177–178
mission, structure, and overview, 173–174, 175f
presentation, 188–189
professional dog training, 188
service animals *vs.* companion animals, 178–180

veteran clients, 174–175
veterinarian wellness exam, 187–188
Pediatric patients
dogs
ASDs, 124–125
hospitalized children, 120–121
mental health disorders, 121–124
reading disorders, 128–129
severe disabilities, 125–128, 128f
elephants, 136
guinea pigs, 135–136
horses
ASDs, 134, 135f
cerebral palsy, 131–134
mental health, 130–131
Posttraumatic stress disorder (PTSD)
canine-assisted therapy, Paws for Heroes
"after-care" financial support, 189
application process, 176–177
assessment process, 181–187
emotional support dog, 189
evaluation and placement process, 181
foster process, 188
ineligible veterans, 177
in-home evaluation, 181
in-person interview meeting, 177–178
mission, structure, and overview, 173–174, 175f
presentation, 188–189
professional dog training, 188
service animals *vs.* companion animals, 178–180
veteran clients, 174–175
veterinarian wellness exam, 187–188
diagnostic criteria, 3, 4t–5t
equine-assisted therapy, 66–72
history, 33–41
history of, 9–16
literature, 33–41
mechanism of action, 9–16
service dogs. *See* Service dogs
sexual trauma. *See* Sexual trauma
signs and symptoms of, 221
treatment
evidence-based psychotherapies, 22–26
patient history, 27
research, 20–21
survivors of trauma, 17–20
Prolonged exposure (PE), 22

R

Reduction, 207
Refinement, 207–209
Replacement, 206–207

S

Safety behavior, 194
Secondary traumatic stress (STS), 55–56
Service dogs (SDs)
challenges in, 56–58
description of, 46–47
family members, 55–56
guidelines, 49–50
indication, 45
mental health benefits, 51–53
physical/medical impacts, 54–55
relational/social/occupational functioning, 54
reported benefits, 50
sexual trauma
disadvantage, 115–116
medical setting, 107–111
strong organization, 111–115
survivors, 102–107
theoretical underpinnings, 58–59
training programs, 47–48
US Department of Veterans Affairs study, 56
Sexual assault (SA), 102–111
Sexual trauma
patient history, 97–102
service dogs
disadvantage, 115–116
medical setting, 107–111
strong organization, 111–115
survivors, 102–107

Shakespeare's, 3, 6–7, 33, 36–37, 221
Stress mitigation, 151–153
Subjective Unit of Discomfort Scale (SUDS), 25–26
Substance abuse disorder, 76

U

US Department of Veterans Affairs study, 56

W

Walking slowly, 202–205, 203f–204f

Kolibri 360 GmbH
Ferdinand-Jühlke-Straße 7
99095 Erfurt, Deutschland
produktsicherheit@kolibri360.de